AFTER THE US SHALE GAS REVOLUTION

FROM THE SAME PUBLISHER

- Heavy Crude Oils
  From Geology to Upgrading. An Overview
  A.Y. HUC

- CO_2 Capture
  Technologies to Reduce Greenhouse Gas Emissions
  J. LECOMTE, P. BROUTIN, E. LEBAS

- A Geoscientist's Guide to Petrophysics
  B. ZINSZNER, F.M. PERRIN

- Acido-Basic Catalysis (2 vols.)
  Application to Refining and Petrochemistry
  C. MARCILLY

- Petroleum Microbiology (2 vols.)
  J.P. VANDECASTEELE

- Oil and Gas Exploration and Production
  Reserves, costs, contracts
  N. BRET-ROUZAUD, J.P. FAVENNEC

- Chemical Reactors
  From Design to Operation
  P. TRAMBOUZE, J.P. EUZEN

- Multiphase Production
  Pipeline Transport, Pumping and Metering
  J. FALCIMAIGNE, S. DECARRE

- Drilling Data Handbook
  G. GABOLDE, J.P. NGUYEN

- The Oil & Gas Engineering Guide
  H. BARON

- Project Management Guide
  M. DUCROS, G. FERNET

All rights reserved.

No part of this publication may be reproduced or transmitted in any form or by any means, electronic or mechanical, including photocopy, recording, or any information storage and retrieval system, without the prior written permission of the publisher.

© Editions Technip, Paris, 2012.

Printed in France

ISBN 978-2-7108-1016-2

▶ **Thierry BROS**
Senior European Gas and LNG Analyst, Société Générale

AFTER THE US SHALE GAS REVOLUTION

2012

t Editions TECHNIP 25 rue Ginoux, 75015 PARIS, FRANCE

The contents of this book are the author's sole responsibility. They do not represent the views of any organizations the author has worked for.

Table of Contents

Preface IX

Introduction XI

26 Centuries of Supply, Demand and Innovation
"Gaz à tous les étages" 2

Basics
Permanent Sovereignty over Natural Resources 3
Reserves 4
    Russia, First Reserve Holder 5
    Iran & Qatar share the Largest Gas Field 6
    Turkmenistan: a Reclusive Oppressive Regime where Information is Scarce 6
Production 7
    Production by Countries 8
    Production by Companies: Gazprom First! 9
Consumption 10
Net Exporters 11
    Geopolitics and Focus on Russia 12
    Cartel Speculation 12
Net Importers 13
    Recent Net Importers: UK and China 15
Trade Movements on the Rise 17

Technicals
Pipe Transport 19
    Transit: Rules 20
    … Issues 20
    …and Solution 20
    China Appears to be emerging as a Key Destination for Gas Supply 22
LNG Chain 24
    Liquefaction 25
    Shipping 31

Regasification	32
Balance between Liquefaction and Regasification	32
From a Quasi-pipe Business to a Grid Business allowing Arbitrages	32
"Dire Straits"	34

Pipe Gas vs. LNG ... 34
- LNG wins in Italy... ... 35
- ...loses in Spain ... 35
- ...and in the UK? ... 37

Uses ... 38
- Historical uses ... 38
- New uses, as a Transportation Fuel ... 39

Gas in Primary Energy Mix ... 40

Demand Seasonality ... 42

Swing Production ... 42

Storage ... 47
- Three Types of Underground Gas Storage... ... 48
- ...and two "Kinds" of Gas! ... 49
- Worldwide Storage Capacity ... 49
- Focus on UK Infrastructure: Storage vs. Imports ... 52
- Russia: Gazprom Monopoly with 65 bcm ... 54
- Ukraine: a Special Position in between Russia and the EU ... 54

Spare Capacity ... 55

Markets, Prices & Costs

Many Price Mechanisms... ... 57

...& Many Units ... 58
- Spot ... 58
- Oil-indexation ... 59

Term Markets ... 60

Costs ... 63
- Cost Structure in Europe ... 63
- Cost Structure in the US ... 64
- Cost of associated Gas ... 65

Europe: from Oil-indexation to spot Markets ... 66
- UK: the Oldest liberalised Market in Europe ... 66
- Future of European Hub ... 69

Markets can mitigate Winter-Summer Spreads ... 73

Markets can Reallocate Supply ... 74
- LNG Arbitrages ... 74

Markets can Select the most Profitable Fuel to generate Electricity ... 79

Markets Integration? ... 81

Policies

Cheap Energy allows Development	83
Secure Energy has been the First Policy Goal as it is Essential to National Security	84
In Oil, Strategic Stocks are a Reality and have been used	84
In Gas, Strategic Stocks are introduced	85
Clean Energy, if we have the Money!	87
EU ETS: too Complex	89
Taxes / Subsidies	91
Regulation / State Intervention	93
Focus on European Storage Regulation	93

Where is the Future Supply Growth?

The US Shale Gas Revolution	97
China holds the Largest Unconventional Gas Reserves	103
China is accessing American Shale Gas Technology	105
… and improving its Gas Infrastructure	106
Production set to Grow by 13% CAGR until 2020e	106
Europe: Incremental Supply in Poland	106
Not in Continental Europe this Side of 2020e	110
Perhaps Offshore in the North Sea	110
Australia could overtake Qatar in LNG, thanks to Unconventional Gas	113
Russia & Norway: a little more Pipe Gas before the Arctic opening	115
Russia: Major Projects under way could Face Challenges	115
Norway has the Flexibility to wield a Degree of Market Power but little Growth left in Traditional Areas	117
Arctic Circle: Barents Sea first	117
Other Places	117
South America	119
South Africa: Shale stopped	119
Mediterranean Sea: Border Disputes likely to delay Israel LNG	120
Caspian Sea	121
Offshore Africa: Unlikely even in Mozambique	121
North Africa	123
Middle East	123

Where is the Future Demand Growth?

In Asia, Mainly in China	125
China Consumption set to Grow by 15.5% CAGR out to 2020e	125
India	126
Japan: 2012e could be the Start of a Plateau Demand	127
Finding New Demand for Gas in North America	128
To generate more Electricity	128

As a Transportation Fuel	128
US Plastics Exports	129
Not in Europe, unless New Uses are Found	129
What about Global Warming?	132
Gas to balance Renewable Energy Production	132
To be Green or Profitable?	132
The Future of Oil-linked Contracts	133
Oil-linked vs. Spot	133
For New Gas Volume to replace Nuclear, Gas has to be priced Differently	136

Conclusion: After the US Shale Gas Revolution, the 2020 Gas World

China Needs to secure Extra Gas from 2018e…	141
… When US LNG could be available!	142
Negatively impacting Russia and Australia	144
Gazprom in the Driving Seat in Europe	147
The Question of Energy Dependency	149
The Remaining Question of European Oil Dependency	150
Units and Conversions	151
Organizations and Data Providers	153
Glossary	157
List of Figures, Maps and Tables	161

Preface

This book gives an in-depth unbiased analysis of today's gas markets, at a time when:

- Asian gas demand is booming thanks to Chinese economic growth and because of the consequences of the Fukushima disaster that have transformed Japan into a nuclear free country.
- US gas industry has to find new demand and build liquefaction trains to mitigate domestic over production due to the shale gas revolution.
- European gas industry has to adapt not only to the third energy package but also to major changes in pricing principle.

After analysing today's gas markets, Thierry is looking at how the different moving parts, not only of the gas supply-demand balance but also of the geopolitics, economics and companies strategy, could shape 2020 gas market. After describing what will be required for shale gas to succeed outside the US, the author concludes that it could help China to slow down its growing dependency and should only have a limited impact in Europe. On the other side, the US would have then become a major LNG exporter. Thanks to those exports, the US could continue to enjoy the cheapest prices until the end of the decade as world gas markets would then be linked via the cost of LNG arbitrages. We could finally see an integration of gas markets with the US Henry Hub price as the worldwide index.

I believe we are facing a period where major gas producers are compelled to move if they want to defend the use of their fuel going forward (displacing oil in transport or coal in power generation).

This book gives the needed tools to better understand those dynamic changes. I therefore hope it can be used as a reference book for people in the gas industry as well as people outside this industry who need to understand it (policy makers, regulators, journalists, students, etc).

Jacques Deyirmendjian
July 1st 2012

Introduction

In the last 5 years we've seen major news flow in the gas business:
- US becoming the number 1 gas producer (since 2009).
- Qatar becoming the number 1 LNG producer (since 2006).
- China becoming a major importer.
- A few transit issues (mainly in Ukraine).
- Major new pipes being built (the 7,000 km Turkmenistan-China pipe and the 1,200 km Nord Stream pipe under the Baltic Sea).
- The largest drop ever recorded in gas demand in Europe (-11% in 2011 vs. 2010) on one side and an unexpected surge in Japan after the Fukushima disaster (+12% in 2011 vs. 2010).

Before analysing the impact and consequences of the most important change (shale gas production in the US), we will first review the current state of the global gas markets. We will then discuss the changes embedded in the US shale gas revolution and the effects on the worldwide supply. We will finally forecast the gas demand for the three major markets (Asia, the US and Europe). For conclusion, we will imagine what the world of gas could look like in 2020.

26 Centuries of Supply, Demand and Innovation

In some places, natural gas is naturally escaping from under the earth's crust and burning. These fires puzzled most early civilizations, and were the root of much myth and superstition. Around 500 B.C., the Chinese first discovered the potential to use this gas to their advantage. They formed pipelines out of bamboo to transport gas where it was seeping to the surface, to cities where it was used for lighting and boiling water. Undoubtedly, it took them many trials and errors (i.e. explosions) to be able to master proper combustion…

In 1799, Phillipe Lebon used gas (produced from coal) to illuminate his house and gardens, and was considering how to light all of Paris… But the first public street lighting with gas was the Pall Mall, in London in 1807. Gas lighting became more common, because towns became much safer places to travel around after gas lamps were installed in the streets, reducing crime rates. At the end of the XIX century, electricity became sufficiently economical and displaced gas as the most popular means of lighting in cities…

In 1855, the Bunsen burner was invented at the University of Heidelberg (Germany). This device that mixed natural gas with air in the right proportions, creating a flame that could be safely used for cooking, heating and sterilisation. The amount of air mixed with the gas stream affects the completeness of the combustion reaction. Less air yields an incomplete and thus cooler reaction, while a gas stream well mixed with air provides a complete reaction with a blue hot flame… that is still used as the logo of some gas companies. The Bunsen burner opened up new opportunities for the use of natural gas. The invention of temperature-regulating thermostatic devices allowed for better use of the heating potential of natural gas, allowing the temperature of the flame to be adjusted and monitored.

"GAZ À TOUS LES ÉTAGES"

Gas displaced coal as easier to use (no heavy bags to carry, no dirty ashes and no unpleasant odor) and new uses appeared: cooking, boiler to heat water (from 1860s) and finally heating (from 1880s). In short, modern comfort at the end of the XIX century was so much gas-fueled that, in France, a plaque "gaz à tous les étages" was sealed on new buildings where all flats had access to gas. A simple way to differentiate comfortable flats and a nice advertisement for gas which can still be seen nowadays!

Photo 1
Gaz à tous les étages.
Source: Collection AFEGAZ/Paris.

From 1960, gas produced from coal and used in Western countries for all of the XIX century was replaced by cleaner natural gas. The gas industry witnessed a major change with local producers, situated on the outskirt of major cities, becoming importers. This book concentrates on the issues the natural gas industry is facing today.

Basics

PERMANENT SOVEREIGNTY OVER NATURAL RESOURCES

Before anything in the energy world where geopolitics, economics and companies strategies interact, it is worth keeping in mind the United Nations (UN) Resolution 1803 (XVII) of 14 December 1962 on permanent sovereignty over natural resources, that declares that:

"The right of peoples and nations to permanent sovereignty over their natural wealth and resources must be exercised in the interest of their national development and of the wellbeing of the people of State concerned.

The exploration, development and disposition of such resources, as well as the import of the foreign capital required for these purposes, should be in conformity with the rules and conditions which the peoples and nations freely consider to be necessary or desirable with regard to the authorization, restriction or prohibition of such activities.

In cases where authorization is granted, the capital imported and the earnings on that capital shall be governed by the terms thereof, by the national legislation in force, and by international law. The profits derived must be shared in the proportions freely agreed upon, in each case, between the investors and the recipient State, due care being taken to ensure that there is no impairment, for any reason, of that State's sovereignty over its natural wealth and resources.

Nationalization, expropriation or requisitioning shall be based on grounds or reasons of public utility, security or the national interest which are recognized as overriding purely individual or private interests, both domestic and foreign. In such cases the owner shall be paid appropriate compensation, in accordance with the rules in force in the State taking such measures in the exercise of its sovereignty and in accordance with international law. In any case where the question of compensation gives rise to a controversy, the national jurisdiction of the State taking such measures shall be exhausted. However, upon agreement by sovereign States and other parties concerned, settlement of the dispute should be made through arbitration or international adjudication.

The free and beneficial exercise of the sovereignty of peoples and nations over their natural resources must be furthered by the mutual respect of States based on their sovereign equality.

International co-operation for the economic development of developing countries, whether in the form of public or private capital investments, exchange of goods and services, technical assistance, or exchange of scientific information, shall be such as to further their independent national development and shall he based upon respect for their sovereignty over their natural wealth and resources.

Violation of the rights of peoples and nations to sovereignty over their natural wealth and resources is contrary to the spirit and principles of the Charter of the United Nations and hinders the development of international cooperation and the maintenance of peace.

Foreign investment agreements freely entered into by or between sovereign States shall be observed in good faith; States and international organizations shall strictly and conscientiously respect the sovereignty of peoples and nations over their natural wealth and resources in accordance with the Charter and the principles set forth in the present resolution."

This resolution, often unknown, sets the rule of the game: State first. Hence why we will look at which states own the gas resources, as they have the right to decide what to do with it (produce it, keep it, export it?). Companies operating in the hydrocarbon upstream sector therefore have an exposure to each of the countries in which they operate.

RESERVES

Resources are hydrocarbons which may or may not be produced in the future. When the relevant government body gives a production licence which enables the field to be developed, reserves can be formally booked.

Reserves are primarily a measure of the probability of gas existing that could be produced. The three categories of reserves generally used are:

- Proven reserves: generally taken to be those quantities that geological and engineering information indicates with "reasonable certainty" can be recovered in the future from known reservoirs under existing economic and operating conditions. Also known in the industry as 1P or having a 90% certainty of being produced.

- Probable reserves: defined as "reasonably probable" of being produced using current technology at current prices, with current commercial terms and government consent. Also known in the industry as 2P (Proven plus Probable) or having a 50% certainty of being produced.

- Possible reserves: defined as "having a chance of being developed under favourable circumstances". Also known in the industry as 3P (Proven plus Probable plus Possible) or having a 10% certainty of being produced.

Oil and gas reserves are the main asset of an Exploration and Production (E&P) company. Booking reserves is done according to a set of rules developed by the Society of Petroleum Engineers (SPE). The reserves of any company listed on the New York Stock Exchange have to be stated to the US Securities and Exchange Commission (SEC). Even if for practical purposes companies and analysts often use proven plus probable reserves (2P), we will stick to proven reserves in this book.

• Russia, First Reserve Holder

According to BP Statistical Review, the top 3 proven reserves holders are: Russia (24%), Iran (16%) and Qatar (14%). So, 3 countries only are holding 53% of the gas proven reserves.

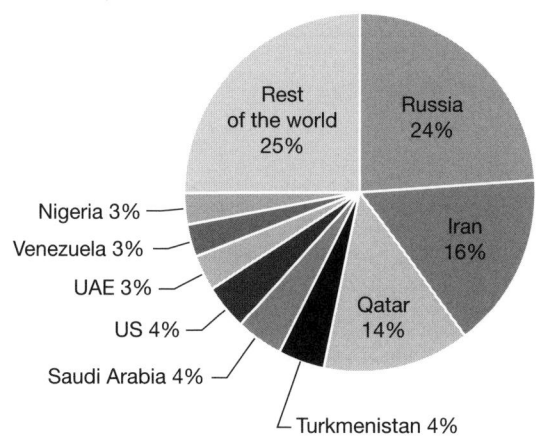

Figure 1

Split of gas proven reserves between major countries at the end of 2010.
Source: BP Statistical Review.

This shows that gas reserves are more concentrated than oil! In oil, the Organization of the Petroleum Exporting Countries (OPEC) controls 77% of the proven reserves but OPEC is an organisation of 12 countries and the biggest oil reserves holder is Saudi Arabia with 19% and the top 3 states controls "only" 44%.

The Gas Exporting Countries Forum (GECF) was set up in 2001, with the objective to increase the level of coordination and strengthen the collaboration between the 12 member States that are gas producers (Algeria, Bolivia, Egypt, Equatorial Guinea, Iran, Libya, Nigeria, Oman, Qatar, Russia, Trinidad and Tobago and Venezuela). The GECF holds 63% of the world's proven reserves of natural gas.

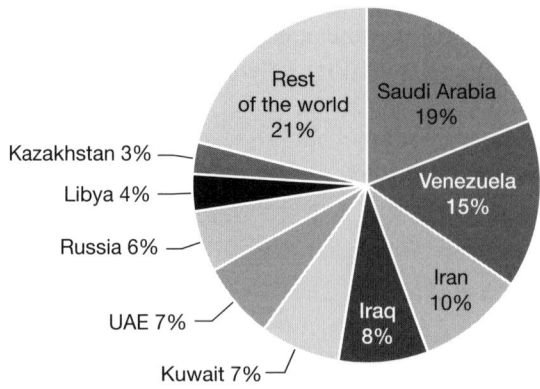

Figure 2

Split of oil proven reserves between major countries at the end of 2010.
Source: BP Statistical Review.

• Iran & Qatar share the Largest Gas Field

After Russia, comes Iran and Iraq. But more than 80% of Qatar's reserves are in the North field, a joint field with Iran, known there as South Pars. The South Pars / North field, located in the Persian Gulf, is the world's largest gas field, shared between Iran and Qatar. This gas field covers an area of 9,700 km^2, of which 3,700 km^2 (South Pars) is in Iranian territorial waters and 6,000 km^2 (North field) is in Qatari territorial waters. Needless to say that this makes production from this field an issue between the 2 States (see production). It is also important to mention that Iran's nuclear program has become the subject of contention with the Western world due to suspicions that Iran could divert the civilian nuclear technology to a weapons program. This has led the UN Security Council, the US and the EU to impose sanctions against Iran, thus furthering its economic isolation on the international scene.

• Turkmenistan: a Reclusive Oppressive Regime where Information is Scarce

Between 2007 and 2008, gas proved resources jumped from 2.6 Tcm to 8 Tcm, thanks to an external audit. But Turkmenistan could perhaps hold the world's second-largest gas field, allowing it to further increase its worldwide share.

Albeit from a very low level, China gas reserves have been rising by 7.5% on a CAGR in 2000-2010. This high growth is laying a solid foundation for the further expansion of chinese gas production.

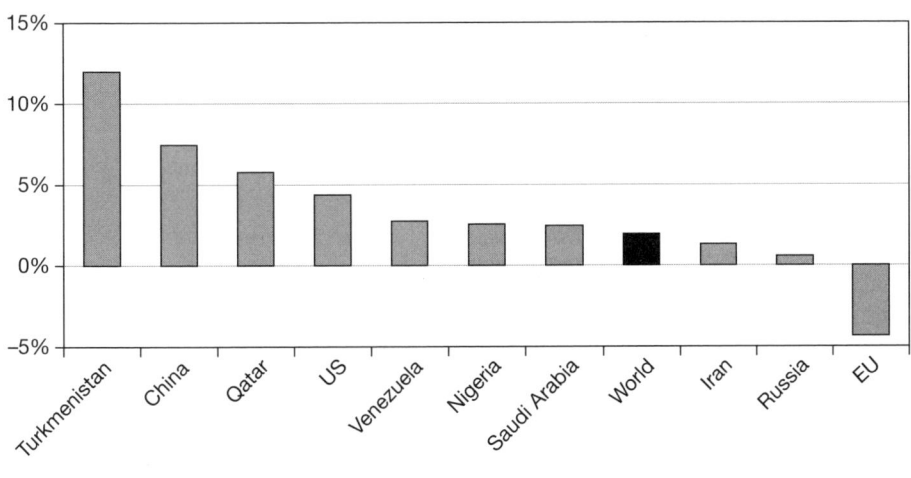

Figure 3

2000-2010 CAGR of gas proven reserves in selected countries.
Source: BP Statistical Review.

Qatar growth in the last 10 years is due to the massive capital expenditure done on the North field to become the biggest LNG exporter.

US growth, twice above worldwide average, comes from the unconventional gas resources that are now been deemed recoverable thanks to innovative technologies.

The EU (and particularly UK with -14% CAGR) saw its gas proven reserves shrink in this 2000-2010 period, as it was producing more gas than it was finding reserves.

For the top 5 reserves holders, the R/P ratio is over 76 years. Then come the US (number 6) where the R/P is "only" 13 years. This is because in the US, private companies are geared towards monetising the resources quickly hence the timing between booking and production is faster than anywhere else. It doesn't mean that in 14 years, the US won't have any more gas reserves because by then, some resources should have been booked into reserves thanks to companies' capex programs. For the EU, the R/P ratio is 14 years but if the EU continues to fail to replace its gas production (and to ban unconventional gas at large), it could mean that in 15 years times, EU domestic production could be insignificant.

PRODUCTION

Total worldwide gas production was 3,193 bcm in 2010 according to the BP Statistical Review.

• **Production by Countries**

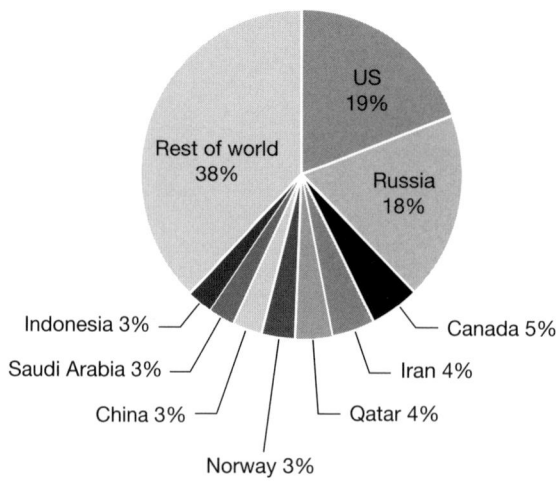

Figure 4

2010 gas production split by countries.
Source: BP Statistical Review.

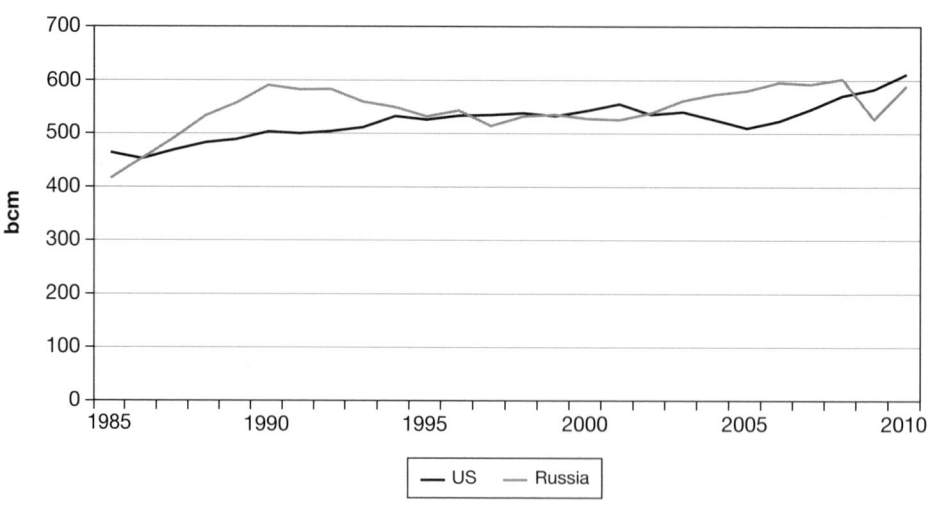

Figure 5

US has overtaken Russia as the first worldwide gas producer since 2009.
Source: BP Statistical Review.

Shale gas was a driving force in helping the US to overtake Russia as the world's largest gas producer since 2009.

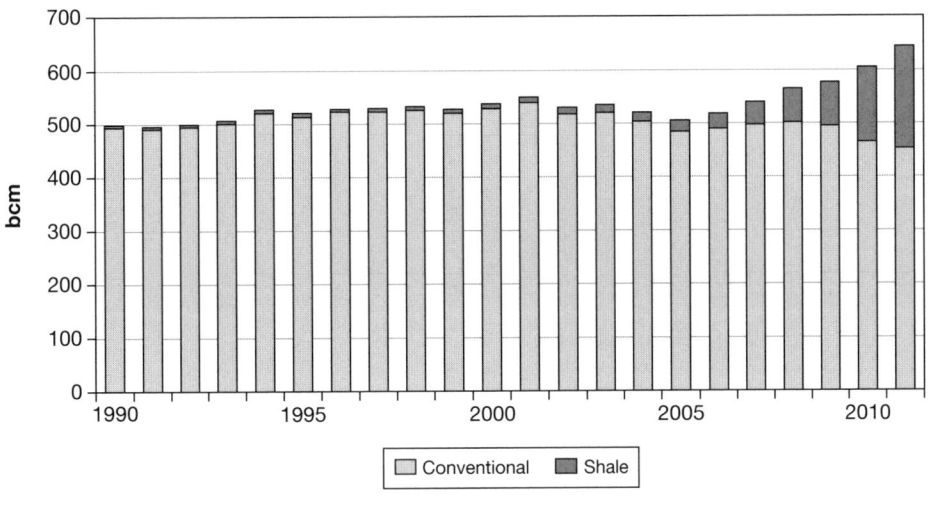

Figure 6

Split of US yearly gas production.
Source: US DoE.

Shale gas production, in the US, increased by a CAGR of 46% over 2005-2010 to account for 23% of 2010 US dry gas production. This is the major change in the gas business. Prior to this, the US production was plateauing and the country was set to become a major gas importer. As we will see all along this book, this shale gas revolution has made the US nearly self sufficient and could potentially allow the US to become a major gas exporter...

• Production by Companies: Gazprom First!

Under the UN Resolution, States have the right to organize gas production as they wish. Hence we have to analyze the worldwide production also on a company basis. Some States have a designated National Oil Company (NOC) that they fully control, some have a controlling stake in their respective NOC and some fully open their upstream to private companies.

The NOCs play by a different set of rules. First, they are subject to a different commercial model because pressure to generate short-term profits is lower as they only have a majority shareholder; the state. While profit is indeed one driver of the behaviour of NOCs, it is just one driver among numerous others. However NOCs have to meet the needs of state budget.

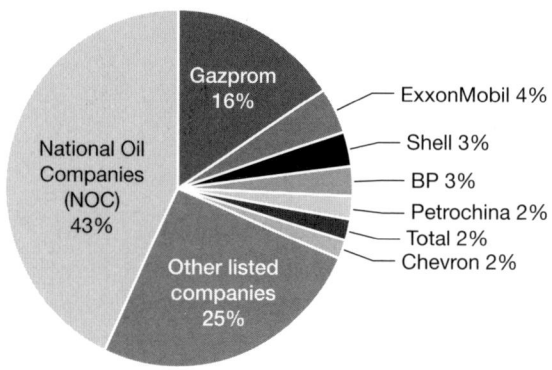

Figure 7

2010 split of production by companies.
Source: SG Cross Asset Research, company data.

As the Russian State allows limited access to hydrocarbon resources, Gazprom (50% controlled by the Russian State) is the major worldwide gas producer. ExxonMobil, which ranks second, still produces only a quarter of what Gazprom produces! ExxonMobil benefitted from the acquisition of XTO in 2010 (for $41bn), consolidating XTO's US unconventional gas production. The first time unconventional gas made the financial news was is in 2005, when ConocoPhillips bought the then independent Burlington Resources for $36bn.

CONSUMPTION

As usual, in all commodities, US are consuming 22% of the worldwide gas…

In the last 10 years, world gas consumption has grown by a 2.8% CAGR with a wide range, from a drop of 3% CAGR in Ukraine to an increase of 15.2% CAGR in China. The way future demand is modeled has a major impact on any forecast. As we will see later, we have to take this wide range into account.

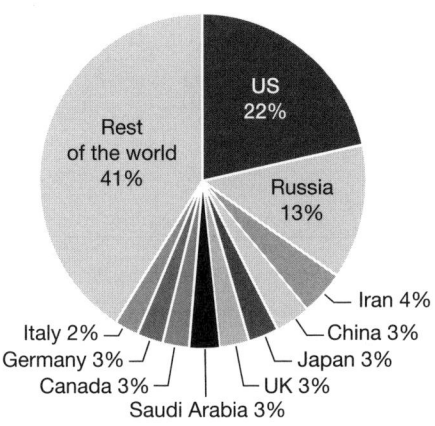

Figure 8

2010 Major gas consumers: split by countries.
Source: BP Statistical Review.

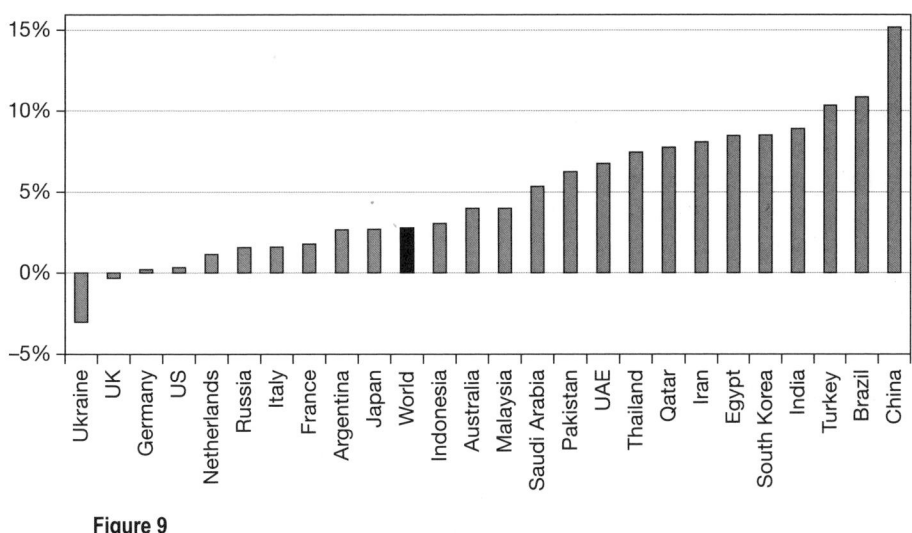

Figure 9

2000-2010 CAGR of gas consumption in selected countries.
Source: BP Statistical Review.

NET EXPORTERS

It is important to see who really exports gas. To assess this we have taken the domestic production minus the domestic consumption as reported by the BP Statistical Review.

Ukraine was the first country in the world to initiate natural gas exports. This was in 1945, with deliveries to Poland.

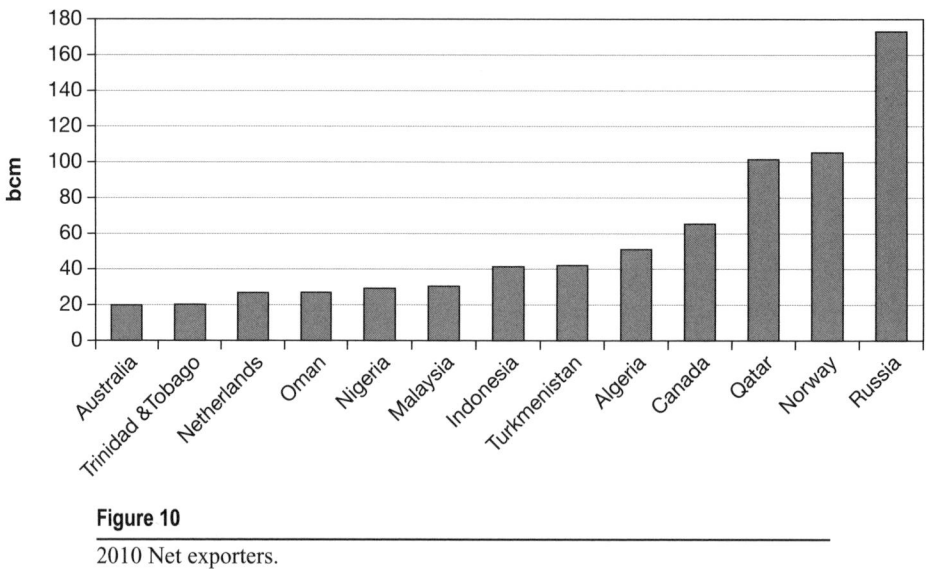

Figure 10
2010 Net exporters.
Source: BP Statistical Review.

• Geopolitics and Focus on Russia

Russia comes out as the major net exporter and Gazprom has been granted the export monopoly for pipe gas. "Can Russia use gas as a political weapon?" is a question often asked. One might also ask "Does Russia need the cash from its gas exports?". It would be better and safer to view the European gas market as a place where Gazprom has a dominant position. No need for a gas cartel...

• Cartel Speculation

GECF is far away from an OPEC style cartel in gas as the organization is still not fully operational and GECF members have little in common: Venezuela is a net gas importer; Iran, due to international sanctions, cannot monetize its vast resources, when all others are net exporters...

On a worldwide basis, GECF has less control in gas over world reserves, world production and exports than OPEC in oil. But, as we will see later, Russia and Qatar have real market power as respectively the major gas pipe exporter and the major LNG exporter.

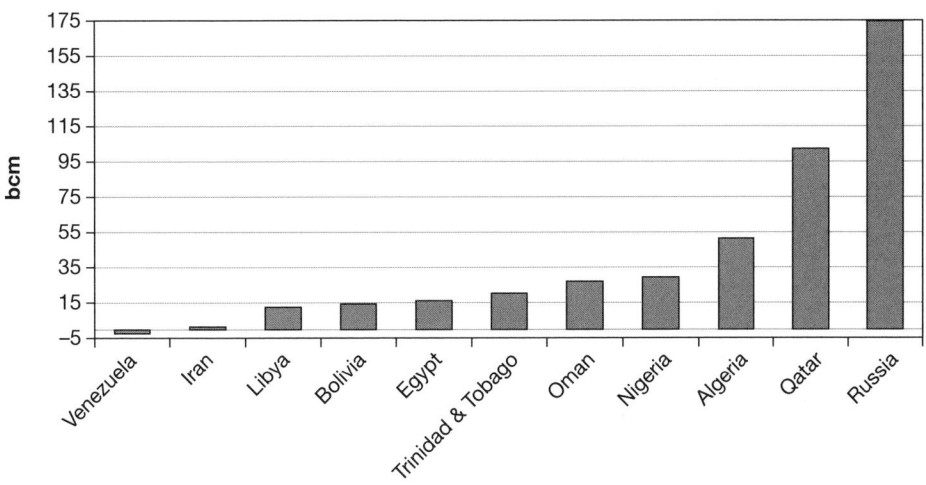

Figure 11

GECF: 2010 net exports.
Source: BP Statistical Review.

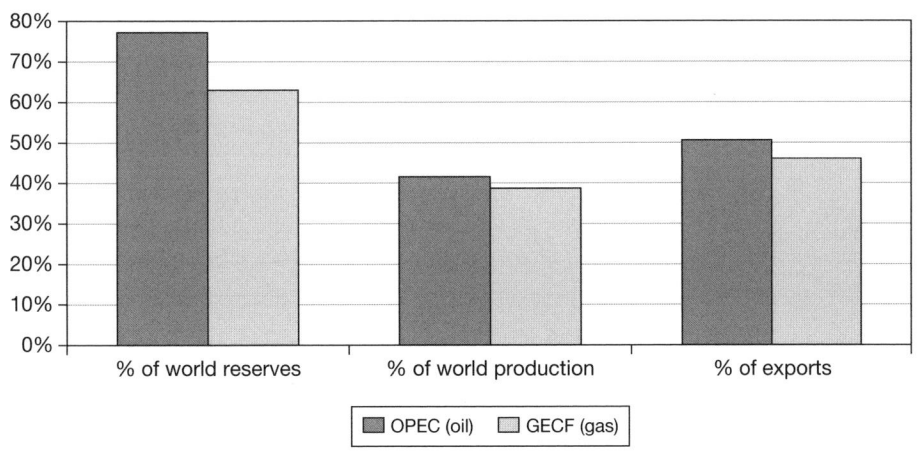

Figure 12

GECF vs. OPEC: OPEC wins!
Source: BP Statistical Review.

NET IMPORTERS

Again, a simplified difference between domestic consumption and domestic production gives the ranking in net importers.

EU collectively is the word largest gas importer with 318 bcm imported in 2010 (and 311 bcm imported in 2011).

But this ranking can change fast due to production and consumption changes. For example, the US net imports have been reduced by 4.7% on a CAGR over the 2000-2010 period thanks to national production growth that comes mainly from unconventional fields.

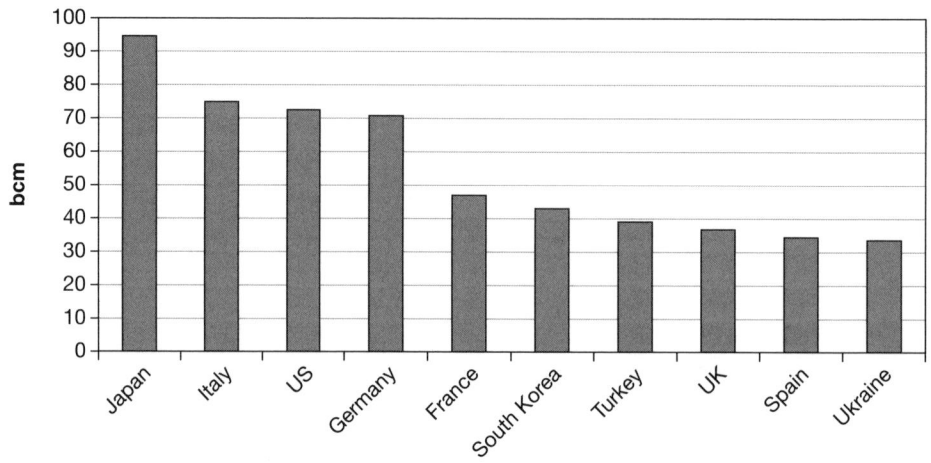

Figure 13

2010 Net importers.
Source: BP Statistical Review.

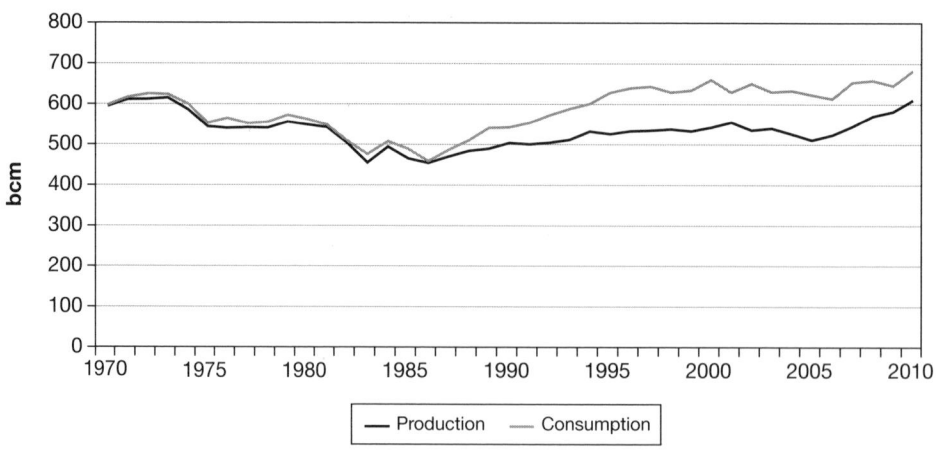

Figure 14

US natural gas production and consumption.
Source: BP Statistical Review.

• Recent Net Importers: UK and China

UK was a net importer of Norwegian gas from 1970 to 1995. The start-up in 1998 of the Interconnector pipeline from Bacton in the UK to Zeebrugge in Belgium resulted in the UK's becoming a net exporter of natural gas for a few years. Both domestic gas production and net exports peaked in 2000. UK production decline sharply since 2000 (–6.2% CAGR in 2000-2010), making the UK a net importer since 2004. And annual UK consumption hit an all-time high in 2004. To balance declining domestic supply and consumption the UK has since 2004 been a pipe gas and LNG importer.

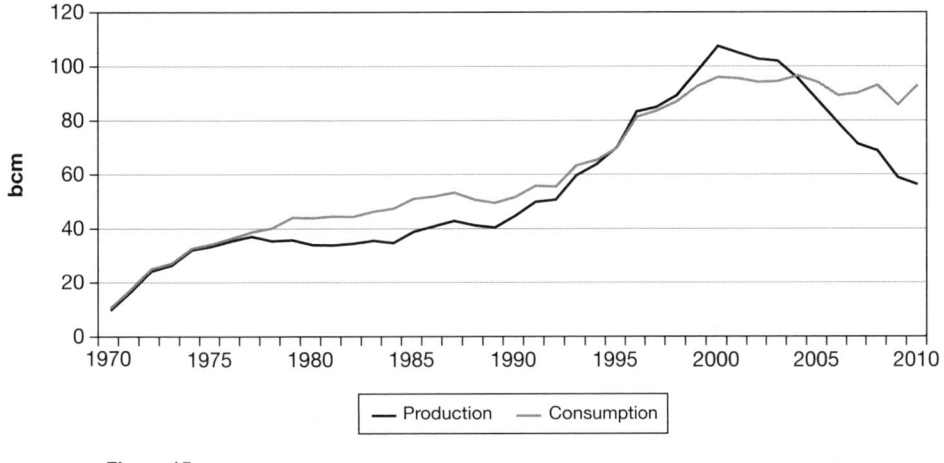

Figure 15

UK natural gas production and consumption.
Source: BP Statistical Review.

China consumption grew strongly in the last decade (+15.2% CAGR in 2000-2010), overtaking the national production in 2003.

China was initially immune to worldwide gas geopolitics but the need to import gas has put China back on the worldwide gas map, 26 centuries after it was the first country to master gas!

We will look at the future of gas production and demand later as we now understand that the metric that matters at state level is the evolution of net imports/exports.

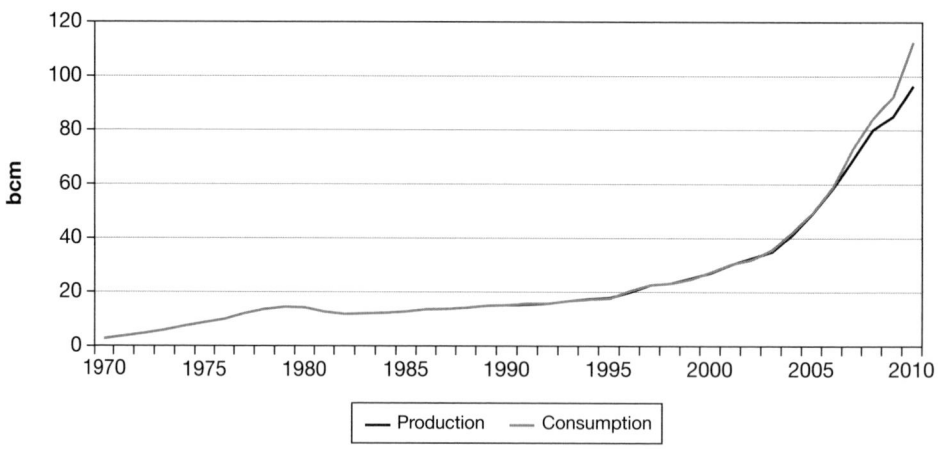

Figure 16

China natural gas production and consumption.
Source: BP Statistical Review.

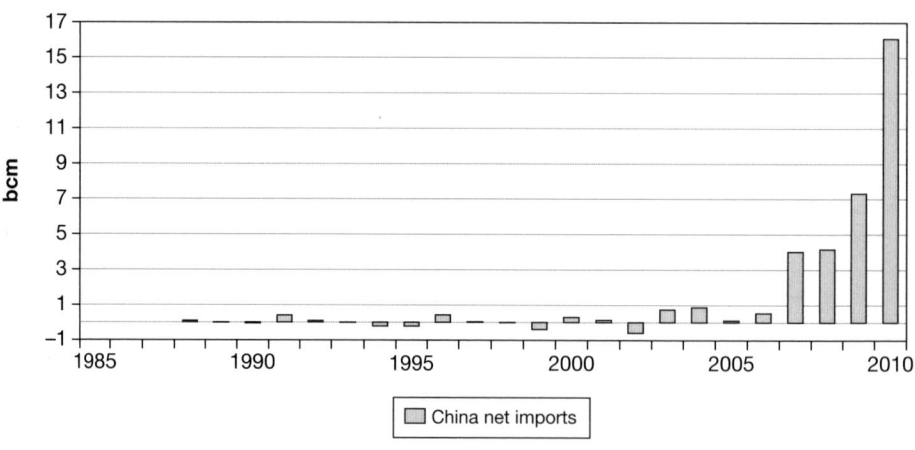

Figure 17

China net gas imports.
Source: BP Statistical Review.

TRADE MOVEMENTS ON THE RISE

According to the BP Statistical Review, total gas imports (except intra Commonwealth of Independent States (CIS) trade) have grown from 554 bcm in 2001 to 975 bcm in 2010. This growth (+6.5% CAGR) is faster than the worldwide production growth over the same period (+2.9% CAGR), allowing total imports to represent 31% of total 2010 consumption.

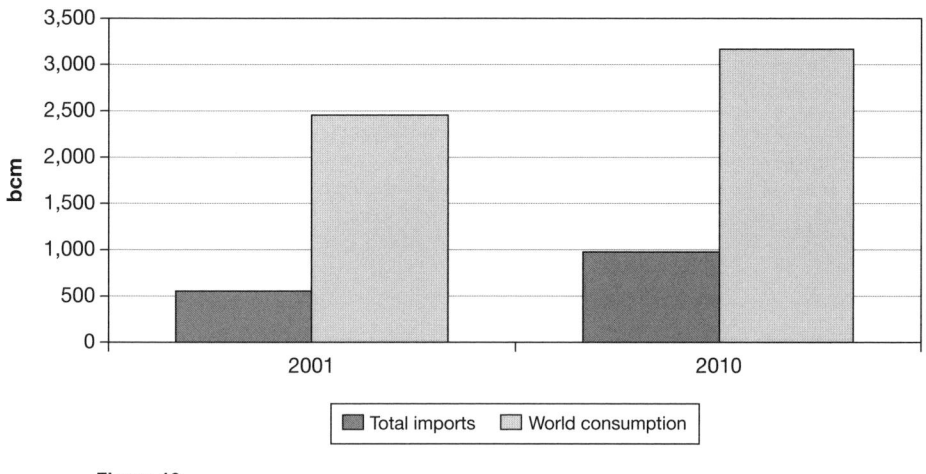

Figure 18

Trade grows faster than world consumption.
Source: BP Statistical Review.

In the last 10 years, the gas world has become more interconnected as consuming countries (like UK or China) needed more and more gas from producing countries. The US countered this trend, becoming nearly self-sufficient thanks to its growing domestic shale gas production. Which trends are we going to witness during this decade?

Technicals

As gas is more costly than oil to transport, gas uses were first developped as close as possible to the production field to avoid flaring. The growth of gas demand was first linked to the gasification of new areas. To develop this consumption away from the producing regions, transport (by pipes or in specific tankers that carry LNG) is needed.

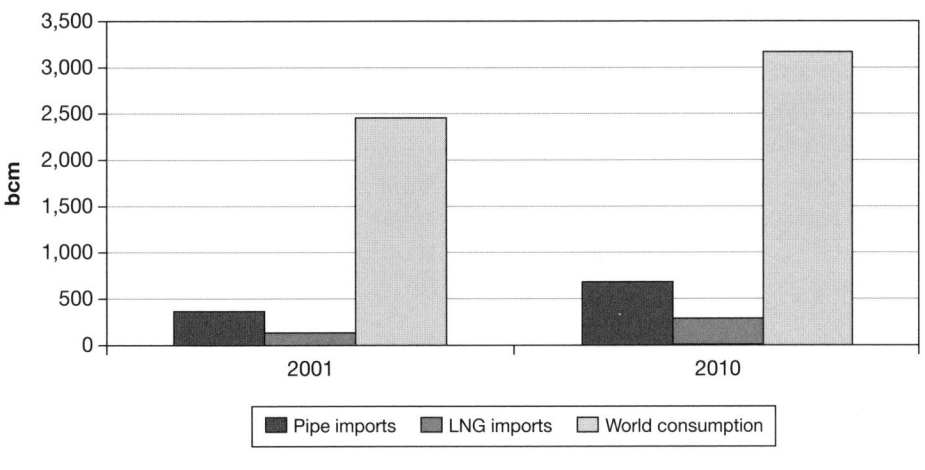

Figure 19
Split between pipe and LNG imports vs. world consumption.
Source: BP Statistical Review.

PIPE TRANSPORT

Pipe is the most common and the oldest way to transport gas from the production field to the customers. Pipes were the first mean to develop this industry thanks to "gasification" of regions further and further away from a gas field. Pipe links the long term producers to the consumers.

• Transit: Rules...

Long distance pipes had historically to be onshore (due to the need of compressors to push the gas) and therefore had to cross different countries before reaching customers.

According to international rules (General Agreement on Tariffs and Trade signed 1946 and the Energy Charter Treaty signed in 1994), freedom of transit is the principle and the traffic in transit is exempted from custom duties and all charges except for transportation and administrative expenses. Transit tariffs shall be objective, reasonable, transparent and non-discriminatory. Tariffs should cover the costs of investment and financing, operating and maintaining the pipe and include an element of profit for the operator. Transiting gas should therefore be on a cost-plus formula, similar to the one used for transporting gas.

• ... Issues...

Quite often in winter, when temperatures are at record lows, Iran cuts exports to Turkey for a few days. This forces then Turkey to cut exports to Greece... But the most well known transit disruptions are the ones affecting Russian gas in Ukraine (2006, 2008 and 2009) and Belarus (2007).

According to the 2009 agreement on terms for supplies of Russian gas to Ukraine and gas transit to Europe, the Ukrainian transit tariff in 2012, should be 3.15 $/1,000 cm/100 km (inclusive of the cost of fuel gas (3%)). The cost of transiting Russian gas in Ukraine should therefore be 39 $/1,000 cm in 2012.

• ...and Solution

Transit issues can only be solved by... avoiding transit!

Only recently, improvement in compressors developed the capability to pipe gas offshore under long distance avoiding transit countries. An offshore pipe (in international waters) is a pipe with no transit issues.

To reduce transit risks, which have materialised in the recent past, Gazprom has bought the Belarus network (Beltrangas is 100% owned by Gazprom since 2011) and is building with its European partners, the Nord Stream pipe. The 1,224 km Nord Stream consists of two parallel lines running from Russia to Germany under the Baltic Sea. Since it began operations in November 2011, the first line of the pipeline has the capacity to transport 27.5 bcm/y of Russian gas to Europe. Once the second line is operational by the end of 2012, Nord Stream should allow Gazprom to export without any transit risk up to 55 bcm/y of gas.

For Nord Stream, taking into account the capex (€7.4bn), the cost of financing and an internal rate of return of 10%, we end up with a price of transit of 2.9 $/1,000 cm/100 km (excluding fuel cost) in 2012. If we take the following assumptions: 400 $/1,000 cm Russian gas cost (to compare it the Ukrainian transit) and a fuel use of 1% (due to increased efficiency of new compressors when the pipe operates at its designed load factor), then transportation via Nord Stream would cost, in total, 39 $/1,000 cm on par with the Ukrainian transit option. For the same price, Gazprom keeps full control of the situation and makes its investment in Nord Stream profitable (and gets 51% of the transportation profit)!

In recent years, Naftogaz of Ukraine provided transit of 95 to 117 bcm/y of gas from East to West (on average 106 bcm/y). With the opening of the Nord Stream (55 bcm/y from 2012e), this could be substantially reduced from 2013e. As there should be only marginal extra Russian volumes contracted in Europe (c. 6 bcm), if we assume a 80% load factor for Nord Stream (44 bcm/y transported), we should see transit volumes via Ukraine being reduced by 38 bcm/y (44-6) to 68 bcm/y. Ukraine should see its transit revenues dropping by $1.5bn (36%) to $2.7bn. To avoid this risk materialising, a minimum amount to ship is stipulated in the 2009 transit contract (110 bcm/y) but no "ship or pay clause" has been included, allowing Gazprom to reduce, if needed, transit volumes without penalty.

Nord Stream, a joint project between Gazprom (51%), Wintershall (15.5%), E.ON (15.5%), Gasunie (9%) and GDF SUEZ (9%), is a strategic investment which enables:

- The partners to mitigate current transit risks associated with Ukraine;

- The partners to provide increased security of supply to their European customers;

- Gazprom to monetise its gas resource in a more secure way, hence increasing the company value.

With the construction of Nord Stream ending (Q4 12e) and the Ukrainian situation not improving, there is now a push, in Russia, to start the construction of South Stream, in 2013e. This should definitively bury the Nabucco project (where gas sourcing has never been secured).

South Stream is to transport gas from Russia across the Black Sea to Bulgaria, from where it would split into two onshore paths, which Gazprom will build jointly with local partners. One line is to run southwest to Greece and into southern Italy. The other will go northwest via Bulgaria, Serbia, Hungary and Slovenia to northern Italy, with an offshoot to Austria. In order to construct the onshore section of the project abroad, intergovernmental agreements have been made with Austria, Bulgaria, Hungary, Greece, Serbia, Slovenia and Croatia. Turkey has agreed for South Stream to transit its maritime territory. If South Stream goes ahead then Ukraine should no longer be a problem as very little gas would transit through it. South

Stream partners (Gazprom (50%), ENI (20%), Wintershall (15%) and EDF (15%)) should take the final investment decision in Q4 12e.

• China Appears to be emerging as a Key Destination for Gas Supply

As China is becoming a major gas importer it needs to build new pipes to access foreign resources. On a geographical level it would make sense to link Chinese consumers to the biggest resource holder (Russia), which happens to share a major border with China... But negotiations for a 68 bcm/y contract have so far not been successful due to different views on pricing.

Eurasian gas is already moving to China. China has been quite successful in laying pipeline and moving Caspian gas to its markets. Under a deal signed in 2006, Turkmenistan undertook to export a total of 30 bcm/y to China for 30 years. An agreement for additional supplies of 10 bcm/y was signed in 2008. An agreement to increase Turkmen gas imports to China to 65 bcm/y was signed in November 2011. The exports of Turkmen gas to China started in December 2009 via the West-East 2 gas pipeline (10 bcm/y capacity). China's West-East 3 gas pipeline will begin operations by the end of 2013e (30 bcm/y capacity). Work on the West-East 4 gas pipeline will begin after 2015e (30 bcm/year capacity).

The majority of the Turkmenistan gas is going to China. It is unlikely that Europe will see, in the foreseeable future, any Turkmen gas, as Turkmenistan is on the East side of the Caspian Sea... a key problem, as the status of the Caspian Sea and the establishment of the water boundaries among the five littoral states are not solved. The Caspian Sea is the biggest worldwide land-locked salted water area. If it is labelled as sea then, according to the 1982 UN convention on the law of the sea, each state should get a share proportional to its Caspian coastline length. If it is labelled as lake then each 5 states should get a 1/5 share of the whole Caspian Sea (this is advantageous to Iran, because it has a proportionally smaller coastline). This disagreement is there to last due to the oil & gas reserves within the Caspian Sea. Numerous oil and natural gas fields lie in disputed areas of the Caspian Sea. Two or more nations claim these fields rendering them virtually untouchable for potential developers awaiting resolution of the situation. But without any agreement it is also impossible to envisage laying any pipe on its sea bed. So even if Turkmenistan is connected to the former Soviet unified gas system supply on the North and Iran on the South, we expect the bulk of Turkmen gas to flow East in the future. A proposed 1,700 km Turkmenistan-Afghanistan-Pakistan-India line (TAPI) with a 33 bcm/y capacity could also, if built, provide gas to India, but this shouldn't materialize this side of 2020e.

China is also building a pipeline from Myanmar. Construction began in 2010 (capacity 12 bcm/y) and the pipe is expected to be operational in 2013e (a long term contract guarantees deliveries of 4 bcm/y).

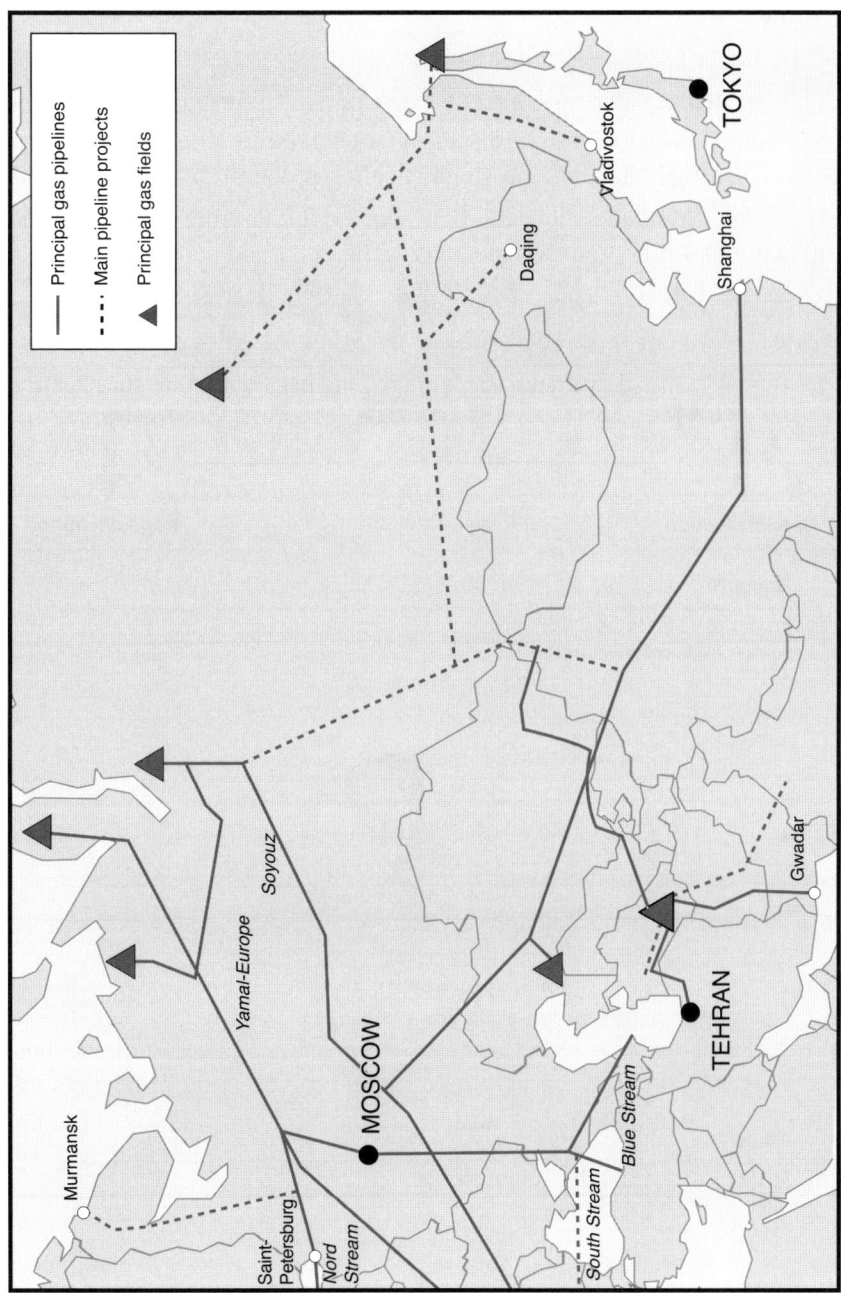

Map 1

From East of Caspian to China

LNG CHAIN

LNG is gas that's cooled down (-162°C) to liquid form (liquefaction process) so it can be shipped by tankers over distances not served by pipelines. Upon arrival, the LNG is converted back into gas (regasification process) for local distribution.

In 1959, the "Methane Pioneer" first shipped LNG from the US to the UK, showing that it could be transported safely, efficiently and economically over long distances by sea. The first LNG trade started, in 1963, between Algeria and the UK.

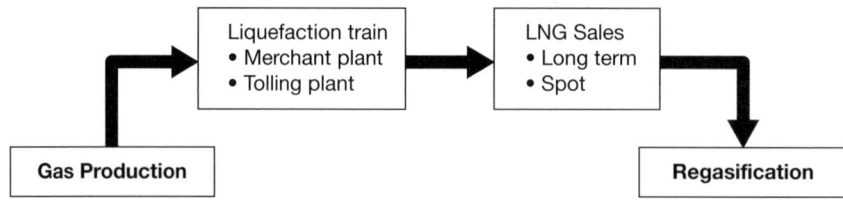

Figure 20

From upstream to final customer.
Source: SG Cross Asset Research.

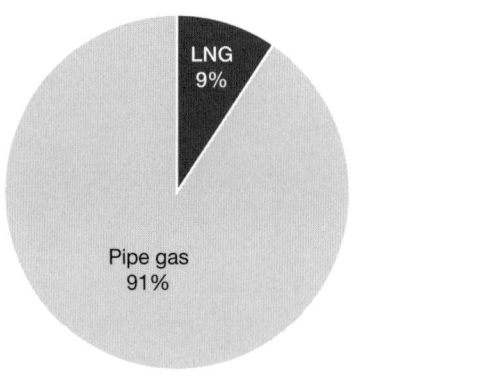

Figure 21

Share of 2010 worldwide gas production.
Source: BP Statistical Review.

If we compare LNG to pipe exports, then LNG accounts for 31% of the total gas exported (except intra-CIS trade).

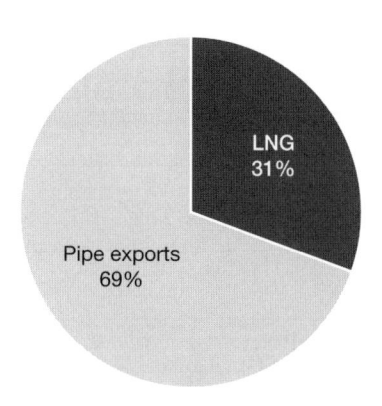

Figure 22

Split between LNG and pipe exports on a worldwide basis (2010).
Source: BP Statistical Review.

• Liquefaction

There were 18 states exporting LNG in 2011. Angola should join them, when its LNG production starts in Q3 12e.

– Qatar undisputed #1...

Qatar, that accounted for 31% of all LNG production in 2011, is the undisputed number 1, but its LNG production (95 bcm) is relatively small compared to Gazprom's production (513 bcm).

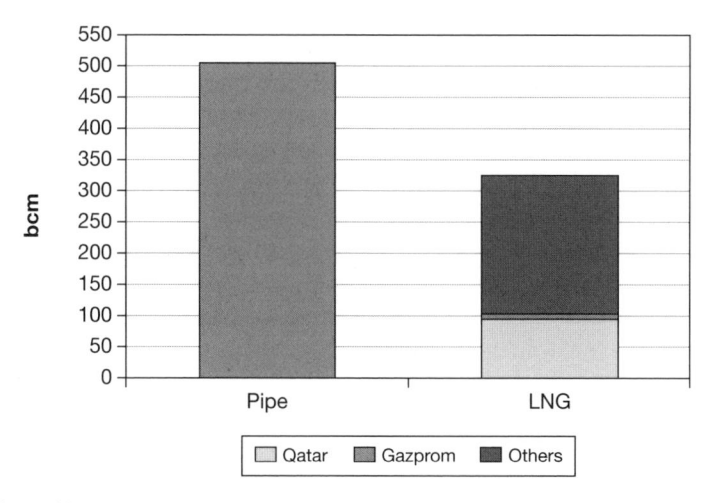

Figure 23

2011 World LNG vs. Gazprom production.

To monetize the gas resource, Qatar Petroleum created joint ventures with major IOCs. But in each of the ventures, Qatar Petroleum has kept a share above 63%.

Table 1 Structure of the Qatari LNG production

Venture	Qatar Petroleum shareholding	Other shareholders	Trains number (and capacity)
Qatargas 1	65%	ExxonMobil (10%), Total (10%), Mitsui (7.5%), Marubeni (7.5%)	Trains 1, 2 and 3 (4.3 bcm/y each)
Qatargas 2	70%	ExxonMobil (30%)	Train 4 (10.5 bcm/y)
	65%	ExxonMobil (18.3%), Total (16.7%)	Train 5 (10.5 bcm/y)
Qatargas 3	68.5%	ConocoPhillips (30%), Mitsui (1.5%)	Train 6 (10.5 bcm/y)
Qatargas 4	70%	Shell (30%)	Train 7 (10.5 bcm/y)
RasGas 1	63%	ExxonMobil (25%), Koras (5%), Itochu (4%), LNG Japan (3%)	Trains 1 and 2 (4.5 bcm/y each)
RasGas 2	70%	ExxonMobil (30%)	Trains 3, 4 and 5 (6.3 bcm/y each)
RasGas 3	70%	ExxonMobil (30%)	Trains 6 and 7 (10.5 bcm/y each)

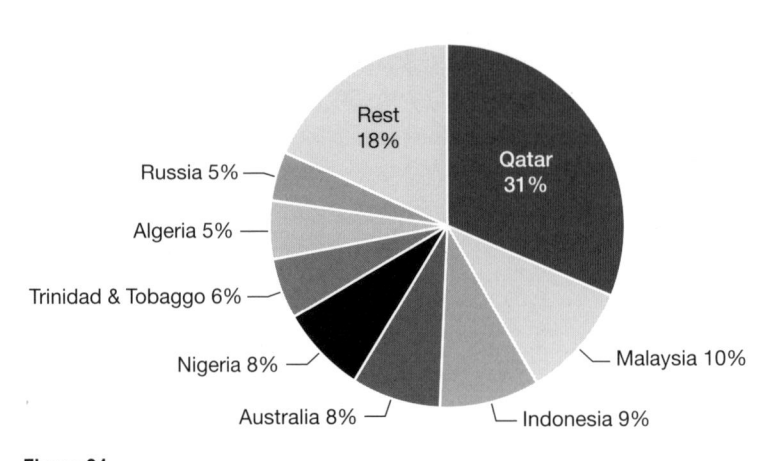

Figure 24

Top 8 LNG exporters in 2011.
Source: GIIGNL.

Qatar, the number one LNG producer (since 2006), which has accounted for more than half of the supply growth in the last seven years, has reached its full export capacity and is not expected to grow in the medium term.

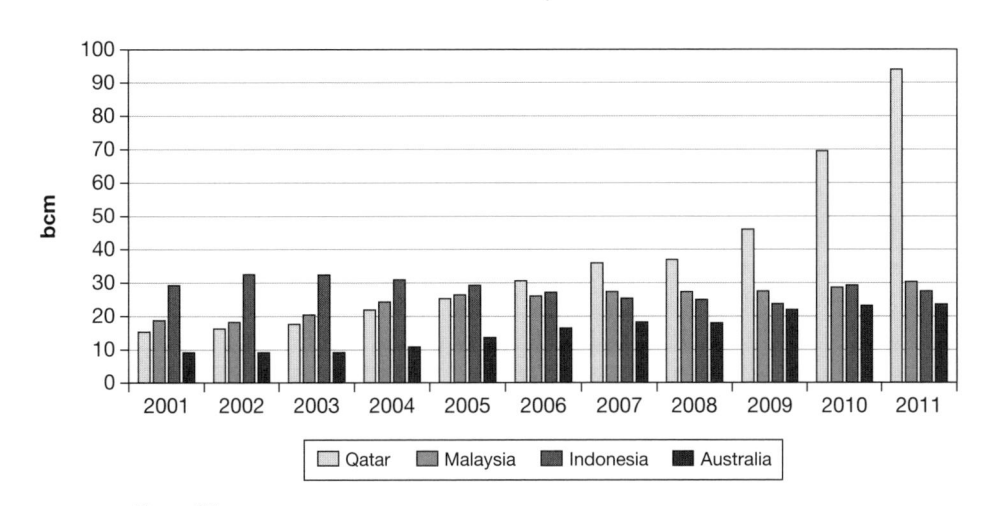

Figure 25

Evolution of LNG production of the 2011 top producers.
Source: GIIGNL.

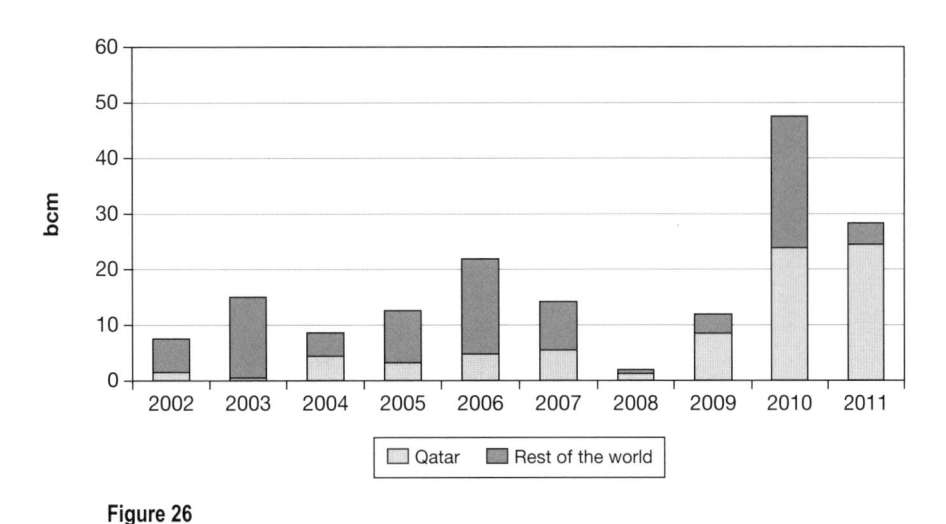

Figure 26

Annual growth in LNG supply.
Source: GIIGNL.

– ... but for how long?

On average, the worldwide liquefaction load factor was 87% in 2011, with a range moving from 8% for Libya (due to the war) to above 100% in Malaysia, Russia and Equatorial Guinea. Civil unrest, increased domestic natural gas consumption and aging gas fields contributed to the weak 2011 production figures for Algeria, Egypt, Indonesia, Oman and Trinidad & Tobago.

The 87% load factor achieved in 2011 leaves little room for improvement in the years to come, as it was a record level for the past seven years. Even if Libya (Marsa el Brega) and US (Kenai in Alaska) come back fully, it is unlikely this load factor will improve as civil unrest and the limitation of LNG exports, as well as routine maintenance, should continue to be a major issue in Algeria, Egypt, Indonesia, Nigeria, Oman and Trinidad & Tobago.

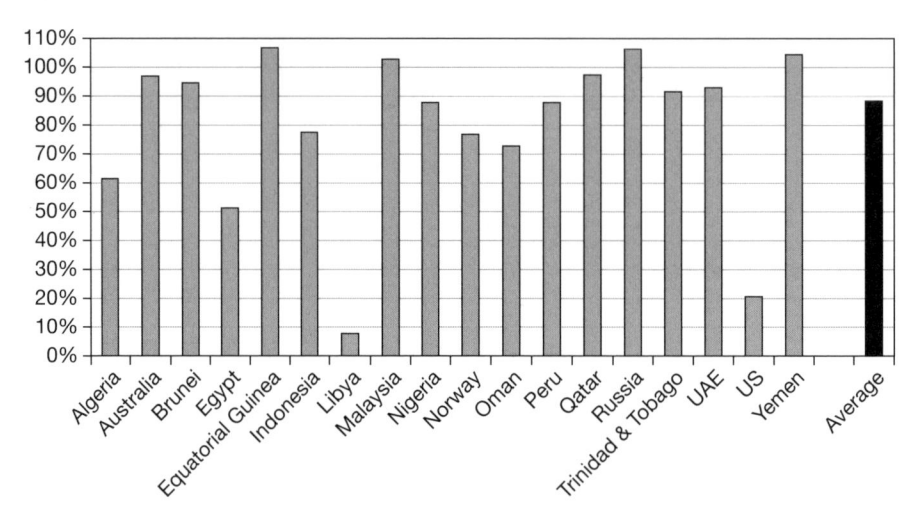

Figure 27

Country LNG load factor in 2011.
Source: SG Cross Asset Research, Waterborne.

We have discounted any growth (new projects or debottlenecking) in Qatar as we believe the moratorium set in 2005 is unlikely to be removed. This is because Iran, which shares the worldwide largest gas reservoir with Qatar would not like to see Qatar monetising this gas resource alone and will not allow any increase in liquefaction capacity before it is also able to produce LNG on its side of the border.

Among all the LNG projects, we highlight the following which could, at best, come online before 2020e (as we are aware that most LNG projects are plagued by delays and cost overruns).

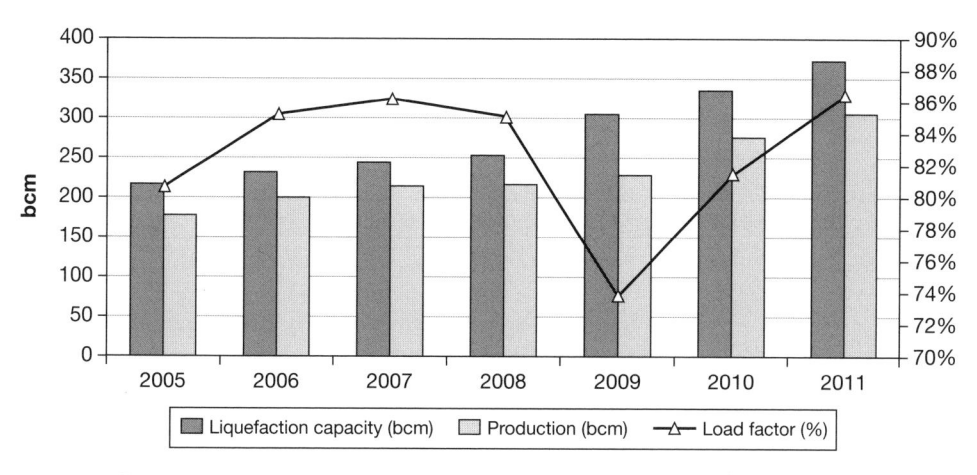

Figure 28
Historical liquefaction load factor.
Source: GIIGNL.

Table 2 LNG projects up to 2020e

Projects	Source of gas	Train no.	Capacity mtpa	Capacity bcm/y	Start-up date estimate
Pluto (Australia)	Conventional	1	4.3	5.8	Q2 12
Angola LNG	Conventional		5.2	7.0	Q3 12
Gorgon (Australia)	Conventional	1	5.0	6.8	2014
Queensland Curtis LNG (Australia)	Unconventional	1 & 2	8.5	11.5	2015
Gorgon (Australia)	Conventional	2	5.0	6.8	2015
Donggi-Senoro (Indonesia)	Conventional		2.0	2.7	2015
PNG LNG (Papua New Guinea)	Conventional	1 & 2	6.6	8.9	2016
Gladstone LNG (Australia)	Unconventional	1 & 2	7.8	10.5	2016
Australia Pacific LNG	Unconventional	1	4.5	6.1	2016

Table 2 LNG projects up to 2020e *(cont.)*

Projects	Source of gas	Train no.	Capacity mtpa	Capacity bcm/y	Start-up date estimate
Sabine Pass (USA)	Hub (both)	1 & 2	9.0	12.2	2016
Wheatstone (Australia)	Conventional	1 & 2	8.9	12.0	2017
Sabine Pass (USA)	Hub (both)	3 & 4	9.0	12.2	2017
Kitimat (Canada)	Hub (both)		5.0	6.8	after 2017
Australia Pacific LNG	Unconventional	2	4.5	6.1	after 2017
Gassi-Touil / Arzew (Algeria)	Conventional		4.7	6.3	after 2017
Gorgon (Australia)	Conventional	3	5.0	6.8	after 2017
Skikda (Algeria)	Conventional	reconstruction	4.5	6.1	after 2017
Marsa el Brega (Libya)	Conventional	reconstruction	0.7	0.9	after 2018
Ichthys (Australia)	Conventional	1 & 2	8.4	11.3	after 2018
Arrow LNG (Australia)	Unconventional	1 & 2	16.0	21.6	after 2018
Yamal LNG (Russia)	Conventional	1, 2 & 3	15.0	20.3	after 2018
Pluto (Australia)	Conventional	2	4.3	5.8	after 2018
Prelude FLNG (Australia)	Conventional	1	3.6	4.9	after 2018
Bonaparte LNG (Australia)	Conventional	1	2.0	2.7	after 2018

Source: SG Cross Asset Research.

The last 2 projects (Prelude FLNG and Bonaparte LNG) will use floating plants to liquefy the gas on site (FLNG for Shell and Floating, Production, Storage and Offloading (FPSO) for GDF SUEZ). Until now, the liquefaction of offshore gas has always involved piping the gas to a land-based plant. Those two projects could, if successful, demonstrate a mean of developing "stranded" offshore gas reserves (those considered uneconomic for development via an onshore plant because they are too small or remote).

Even assuming Pluto and Angola can produce at full capacity in 2013e, this leaves very little room for capacity growth until 2014e.

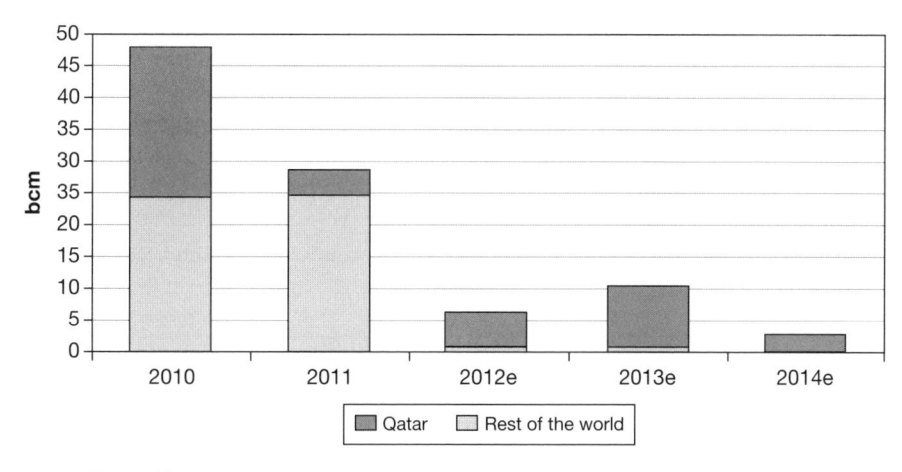

Figure 29

New liquefaction capacity 2010-2014e.
Source: SG Cross Asset Research, GIIGNL.

With 6.3 bcm/y in contracts that are ending between 2013e and 2015e (for Asian customers), we can assume that Indonesia, that is facing production issues, could see a decline of 5.4 bcm/y from 2011 level in 2015e and a production plateauing at 23 bcm/y until the end of the decade.

• Shipping

At the beginning of 2012, 359 ships could carry LNG with an additional 68 on order. Vessel capacity runs from 18,800 cm to 265,000 cm of LNG with an average size of 150,000 cm. The difficulty of the LNG business is to have a balanced market in liquefaction / shipping and regas. Any sudden change in supply (disruption) or in demand (like the nuclear phase out) creates relaxation or tightness in the shipping industry.

LNG is also distributed by very small tankers (1,000 cm) to small islands. But this very small business is outside the scope of this book.

• **Regasification**

The UK was the first country to import LNG, in 1963. With the North Sea discoveries, the UK then stopped importing LNG from 1979 until 2005, when the Isle of Grain regas terminal went online. India's arrival on the LNG scene in 2004, was followed by China (2006), Mexico (2006), Argentina (2008), Brazil (2008), Chile (2009), Kuwait (2009), Dubai (2010), Netherlands (2011), Taiwan (2011) and Indonesia (2012). Malaysia (that like Indonesia is an LNG exporter) should be the 27th country to join the LNG importer club in 2012e.

Since 2006, Chinese LNG imports, driven by strong economic growth and power demand, have grown by 78% on a CAGR. With receiving 5% of the global LNG in 2011, China's growing presence on the LNG scene is likely to attract more and more attention as it has already secured 46 bcm/y of LNG under long term contracts.

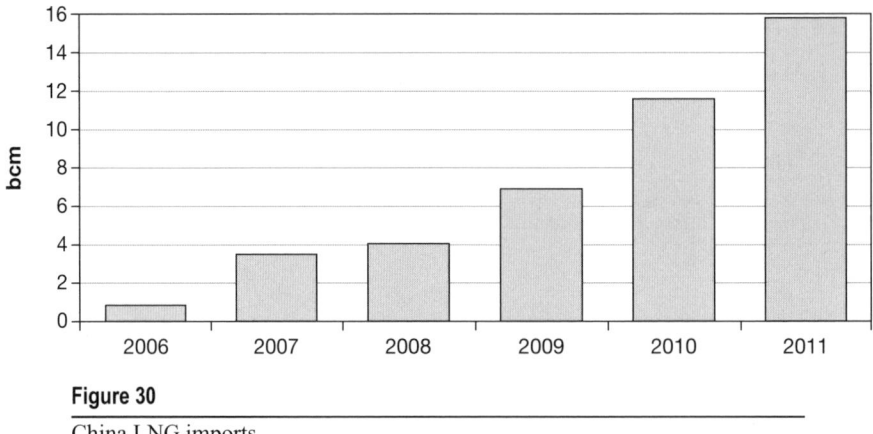

Figure 30

China LNG imports.
Source: GIIGNL.

• **Balance between Liquefaction and Regasification**

A liquefaction plant costs a minimum of $5bn (and projects can go above $35bn), whereas a regasification plant cost less than $1bn. Due to the difference in capex, the regas capacity/production ration was 2.3 in 2011 on a worldwide basis, implying a theoretical load factor of 35% for all regas capacity.

• **From a Quasi-pipe Business to a Grid Business allowing Arbitrages**

Since the beginning, in 1963, long-term contracts (20-year) have been the backbone of this industry. The entire LNG chain was built at the same time: a liquefac-

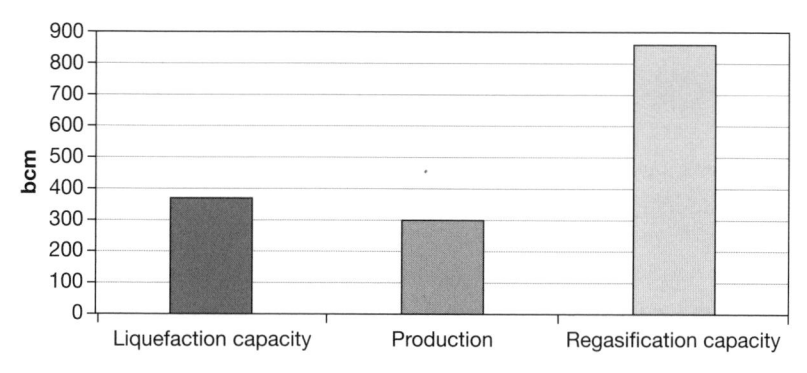

Figure 31

Worldwide liquefaction capacity and 2011 production vs. regas capacity.
Source: GIIGNL.

tion plant in the producing country, ships in yards and a regasification plant in the consuming country.

Arbitrages were previously infrequent but at the turn of the millennium, some companies started to implement a more lucrative business model by redirecting more and more cargoes as rerouting of a cargo could bring huge profit. This flexibility also helped rerouting cargoes away of the US (since 2010 and the shale gas boom) and into Asia after the March 2011 Fukushima disaster. We will look further into this once we explained the different gas pricing mechanisms.

The ranking of top LNG importers has changed rapidly in the past and could change even more due to the flexibility of the LNG chain.

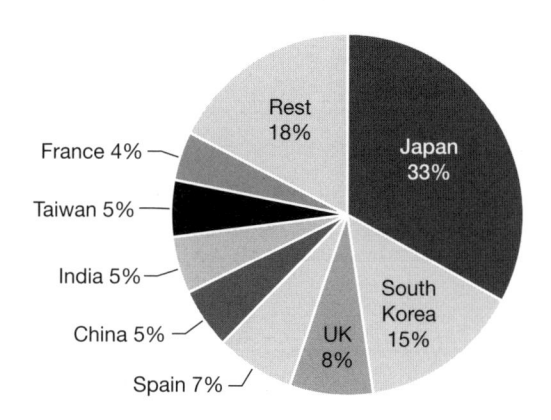

Figure 32

Top 8 LNG importers in 2011.
Source: GIIGNL.

• "Dire Straits"

Even if LNG doesn't face transit issues as pipe gas does, it needs to pass through straits or canals that are chokepoints.

The closure of the Strait of Hormuz is the risk most often considered by the market as Israeli leaders could favour unilateral military action to slow Iran's pursuit of a nuclear weapon. Even if 34% of the world's LNG passes through the Strait (33% outflow from Qatar and the UAE and 1% inflow to Kuwait), we argue that this risk is small as the blockade could not be effective for very long, thanks to international marine intervention. Iran's rhetoric is tough even against other OPEC producers, warning them not to increase their oil production in the event that it cuts off exports to the EU. Thus we think a higher risk could be the possibility of Iran bombing the Ras Laffan Qatari LNG production complex (Kuwaiti oil overproduction has been a serious point of contention with Iraq in the past, prompting Iraqi military forces to set fire to Kuwaiti oilfields in January-February 1991). Such a strike could be "easily" carried out by launching missiles from Iranian navy ships cruising in the area. The UN Security Council would likely qualify this as an act of war. But for Iran, it could be retaliation against Qatar for too rapidly depleting the North field/South Pars reservoir that the two countries share… Without elaborating on the political repercussions, the consequences on the gas scene would be huge as, on the day of such an attack, 104 bcm/y capacity (30% of the worldwide LNG capacity) would be totally destroyed and could not be put back on line for at least six years. Such action would lead to all un-contracted LNG being rerouted to Asia. But as uncontracted LNG accounts for only around 25% of the LNG supply, this would not be enough. In this case, we would see not only the UK but also continental Europe receiving virtually no LNG, as all the supply would be going to Asia. This would allow Gazprom to provide more pipe gas volumes. We believe that the European system could just cope with such a stressed situation, thanks to Gazprom's spare production capacity (that we estimate at 70 bcm/y) and the new transport route (Nord Stream), as we will see later.

Other choke points are the Malacca straits, the Suez Canal and the Panama Canal when enlarged, in 2014e.

PIPE GAS vs. LNG

In a world where interconnections are improving, pipe gas and LNG should be in competition. We analyse three recent European cases.

The EU has LNG access thanks to Belgium, the UK, France, Italy, Greece, Portugal, Spain and the Netherlands. The total actual regas capacity for the EU is 178 bcm/y.

Figure 33

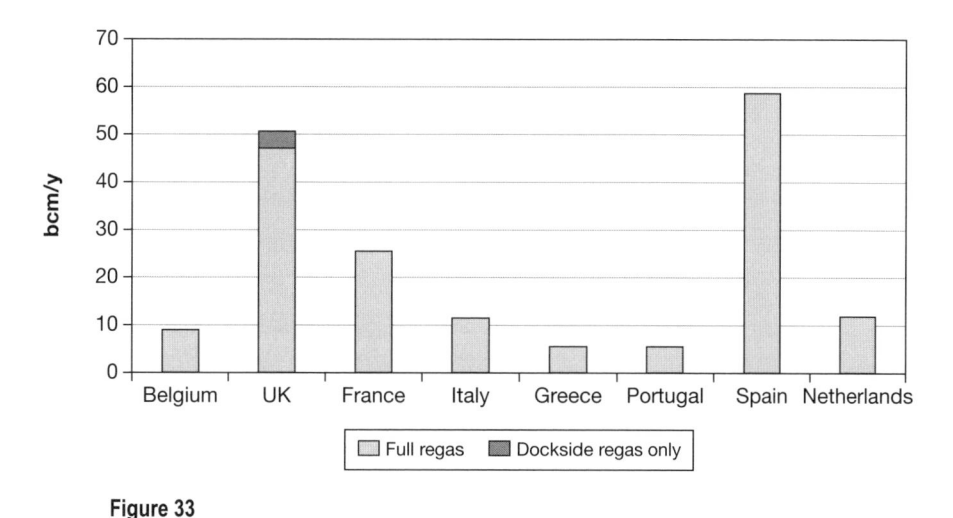

European regas capacity by country.
Source: GIE.

As US regas terminals are not used much (due to the shale gas revolution that has made the US nearly self-sufficient), the average load factor in 2011 for the EU was 46%, with a minimum of 6% for Netherlands (as the terminal opened late in 2011) and a maximum of 75% in Italy.

• LNG wins in Italy...

The opening of the Adriatic LNG in Italy in 2009 (8 bcm/y capacity) has allowed Qatari LNG to displace Russian pipe gas since then. In 2010, when Italian demand increased by 6.3% (vs. 2009), the Adriatic LNG operated at 89%, taking most of the consumption growth. Pipe imports didn't benefit from the growth in Italian gas demand...

This Qatari LNG forced ENI, the major Italian gas buyer, to take less Russian pipe gas than its contractual minimum as we will see later...

• ...loses in Spain...

In LNG, Spain lost its position as the third biggest market in 2011 after Japan and South Korea. The UK (23 bcm imports in 2011, +30% vs. 2010) has overtaken Spain (22 bcm, -16% vs. 2010) as the biggest LNG buyer in Europe. Spain's LNG decline in 2011 vs. 2010 can be attributed to: 1) the stop of the Libyan LNG exports (0.3 bcm in 2010); 2) the opening of the Medgaz line that now allows more pipe gas from Algeria to reach Spain and 3) the decline in gas demand in Spain.

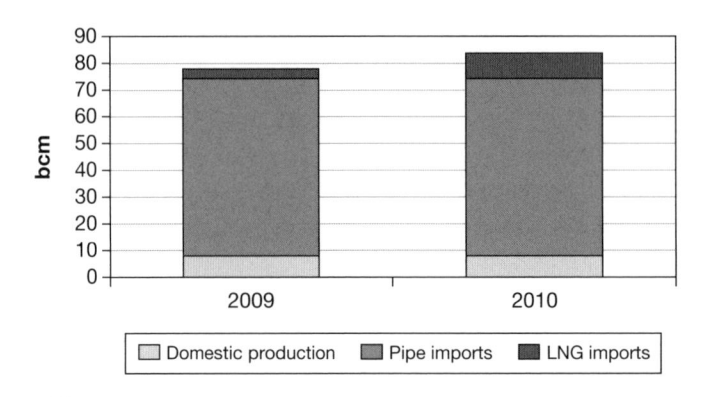

Figure 34

Italian split of gas demand after the opening of a new regas terminal.
Source: SG Cross Asset Research.

Since starting operation in March 2011, Medgaz has slowly ramped up and Algerian pipe gas has and should continue to displace LNG in Spain.

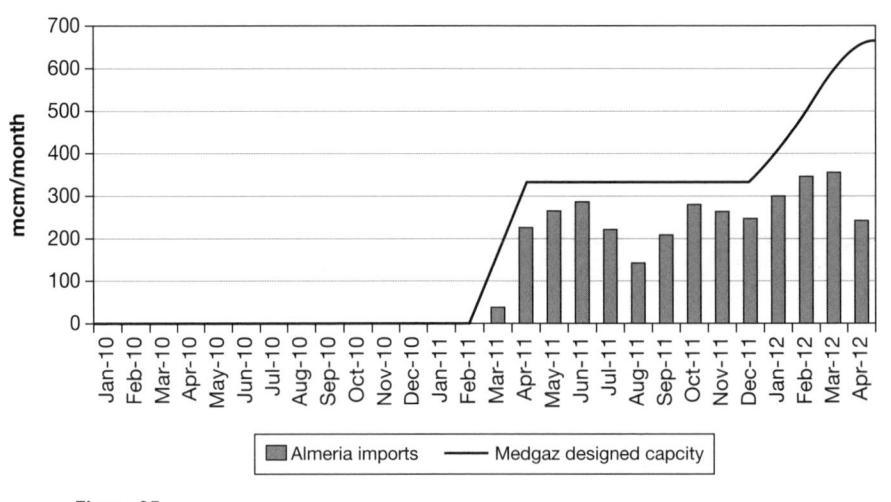

Figure 35

Medgaz should further displace LNG out of Spain.
Source: SG Cross Asset Research, Enagas.

For 2011, with Spanish gas demand down by 0.3 bcm (vs. 2010), LNG imports were reduced by 3.9 bcm to accommodate for an extra 2.1 bcm pipe gas that arrived through the new Medgaz pipe.

Figure 36

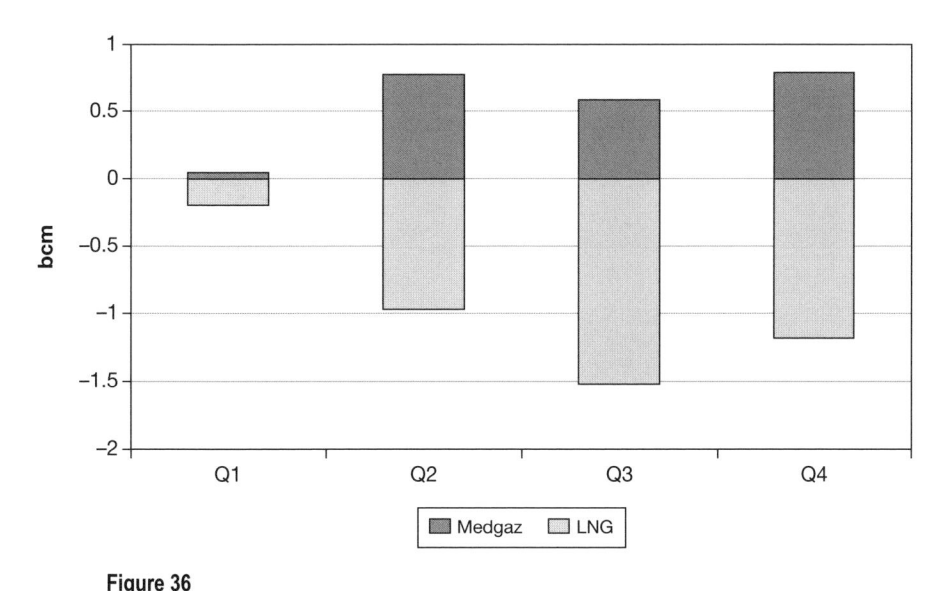

2011 vs. 2010 quarterly variations of Medgaz pipe gas and LNG imports in Spain.
Source: SG Cross Asset Research.

This competition between pipe gas and LNG explains the announcement, in April 2012, that the El Musel LNG receiving terminal will not be put into service upon completion of construction later this year. The 6.9 bcm/y capacity Spanish terminal, which has been under development since 2006, will be mothballed.

• **...and in the UK?**

The UK has seen a major increase in LNG import capacity with the opening of 4 regas terminals since 2005 for a total capacity of 51 bcm/y (for a UK demand of 90 bcm/y). The uncertainty is how such capacity will be utilised and how the UK will compete for LNG on a global basis.

Because of the low nuclear production in Japan in 2012, the diversion of cargos to Asia, that reduced the UK's LNG imports in the second half of 2011, is now accelerating. Since January 2012, Spain's LNG imports were once again higher than those of the UK. In an extreme scenario we could see the UK with very little LNG. In this case, extra Russian pipe gas could then be called in to balance demand.

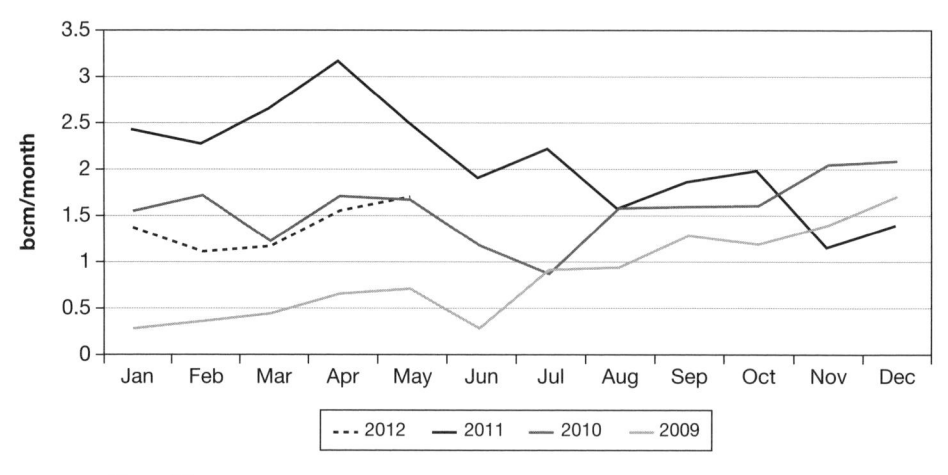

Figure 37

UK LNG imports.
Source: SG Cross Asset Research, Waterborne.

USES

• Historical uses

Once gas reaches a consuming country (by pipes or LNG) it moves through a transmission system at 30 km/h, which means that additional supplies cannot be delivered instantaneously when demand increases. Far away supply cannot instantly arrive at the location of increased demand so the gas transmission system has to be kept constantly in balance.

The high pressure part of the transmission system supplies gas to power stations, a small number of large industrial consumers and to local distribution networks that contain pipes operating at lower pressure, which ultimately supply residential consumers.

Gas uses are split in 3 major categories:

– *Residential*

The level of residential (heating and cooking) and commercial gas demand are largely influenced by temperature on a short-term basis. In OECD countries, energy efficiency measures are reducing, on the long term, the volume of gas needed for heating houses.

– *Industrial*

The industrial demand growth is mainly correlated with Gross Domestic Product (GDP). This explains why gas demand is now booming in non-OECD countries.

Energy intensive industrial sectors are very price sensitive and, in the last 10 years, some fertilizers and petrochemical plants have moved away from OECD countries to producing countries (Middle East) where the price of gas, a by-product of oil production, was very competitive. We will explain, after we looked into prices, why this trend could shift in favour of the US.

– Power generation

Gas is burned in gas-fired power station to produce electricity. The power sector remains the most variable as the fuel mix (renewable, petroleum products, gas, coal and nuclear) can be altered due to one fuel becoming more competitive or due to regulation (in favour of renewables in Germany, in favour of coal in Spain and against nuclear in Germany and Japan, just to name a few). We will discuss inter-fuel substitution and competition more in-depth once we have a better view of prices. In short, the merit order to generate electricity is first solar and wind when available, baseload nuclear, cheap hydro, and then gas, oil or coal depending of their respective pricing. Those different inputs for power generation allow fuel competition to take place as we will see later. Thanks to the high flexibility of modern gas and coal-fired power stations, gas and coal are the back-up fuels able to provide electricity when intermittent solar and wind are not available.

We will also look at the consequences of the Fukushima disaster as nuclear free Japan needed more gas (LNG), coal and oil to avoid serious blackouts.

On average, the split is c. 1/3 for each sector.

• New uses, as a Transportation Fuel

In 1991, LNG experimental locomotives were developed in the US.

In 2000, LNG was first used to power non LNG tankers, with more than 30 LNG-fuelled vessels in operation today. The adoption on the Regulations for the Prevention of Air Pollution from Ships included in Annex VI of the International Convention for the Prevention of Pollution from Ships (MARPOL) should promote LNG-fuelled vessels in the future as fuel used to propel ships should see their sulphur content being reduced in the future. Bunker fuel will have to be phased out in the years to come and LNG could take this market as it is much cheaper than gasoil.

In 2005, natural gas fuelled drilling rigs began to be used in the US, followed in 2010 by the first US drilling rig using LNG as a primary fuel. US gas companies should, in the years to come, favour those rigs against the old diesel powered ones.

Today, 61% of the vehicles on the road in Bangladesh are natural gas powered. In Pakistan, this ratio is 37%, but less than 1% in the US. This means that the US has an enormous room for growth. This could materialise faster than expected thanks to gas prices much lower than oil.

In 2010, the first LNG mining truck was used in the US.

In France, today 15% of the buses are powered with Compressed Natural Gas (CNG) and trials started, in 2011, for LNG powered trucks. With a total of 7,000 gas powered vehicles, the annual consumption is 0,03 bcm/y (or less than 0.1% of total French consumption).

Even if this demand seems negligible today, this could be the only growth area for gas...

GAS IN PRIMARY ENERGY MIX

The amount of gas in the primary energy mix varies from country to country. On a worldwide basis, gas represented, in 2010, 23.8% of the primary energy mix, with a record of 90% in Trinidad & Tobago. "Gasification" started close to the production fields, hence the high use of gas in energy mix in producing states and transit states, with the exception of Norway that produces its gas for other European countries as it can rely on hydro for 64% of its primary energy mix.

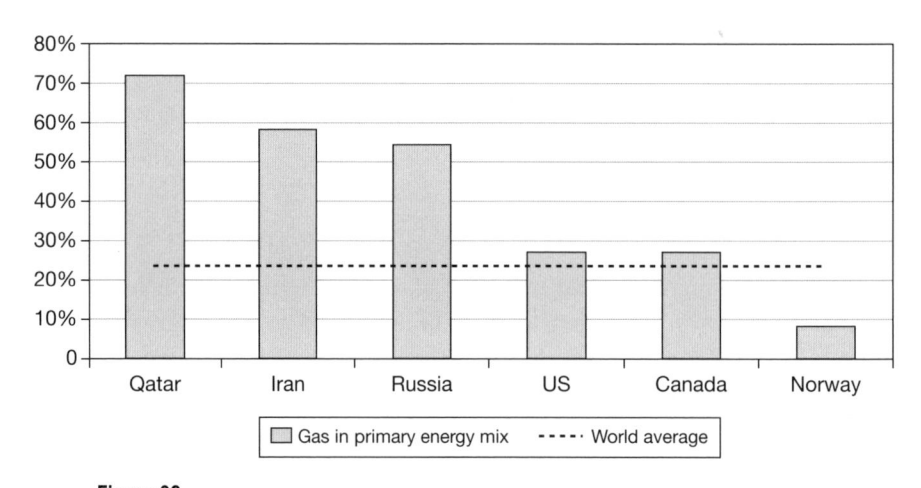

Figure 38

Gas in primary energy mix for major gas producers.
Source: BP Statistical Review.

China is one of the major consumers relying the least on gas in its energy mix (only 4%) but it aims to increase it to 10% by 2020e to curb pollution.

Conversely, in the EU this split varies widely from 3% in Sweden to 42% in Hungary, with an average of 26%. This average could move up as Europe wants to turn away from oil in favor of greener fuels.

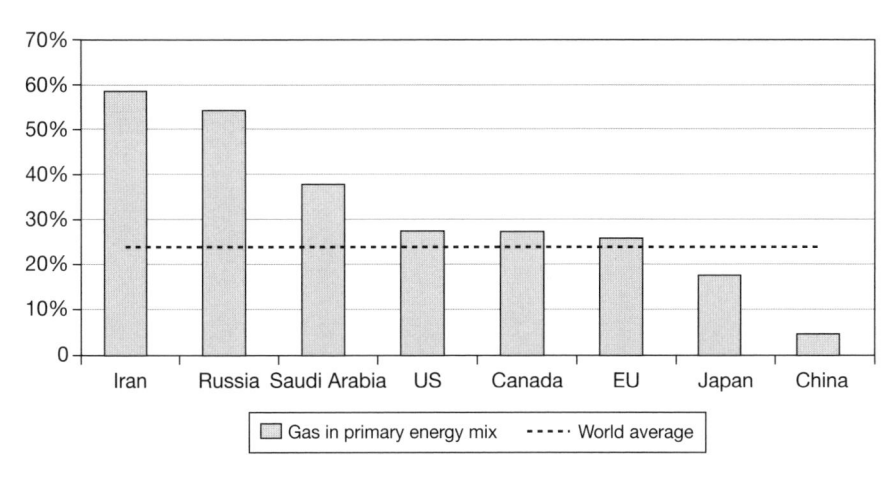

Figure 39

Gas in primary energy mix for major gas consumers.
Source: BP Statistical Review.

The high use of gas by transit states is linked to history. When the pipes were built those transit countries got discounted gas prices. This discount in gas price tends to be challenged with time by the producer... leading to a transit crisis.

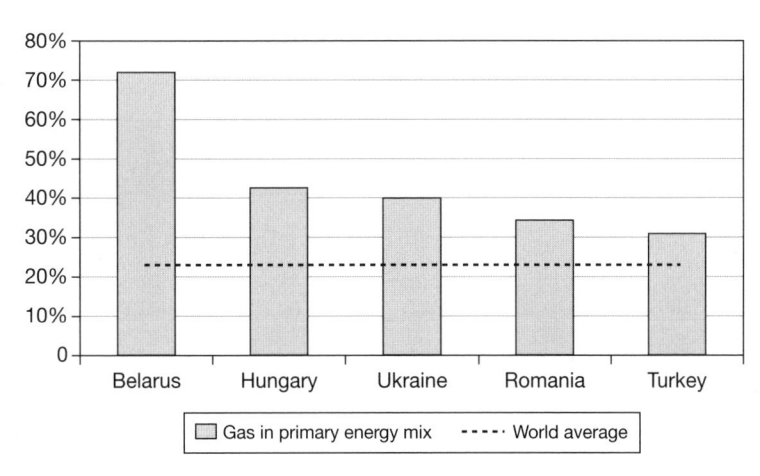

Figure 40

Gas in primary energy mix for major transit states.
Source: BP Statistical Review.

DEMAND SEASONALITY

To gauge demand seasonality we divided the month when demand was as it highest by the month when demand was at its lowest.

As monthly gas data is only available for OECD countries, this metric gives on average 1.8 with a maximum in Europe (2.7) and a minimum in Mexico (1.2). For the biggest non-OECD gas consumer (Russia), this metric is 2.5.

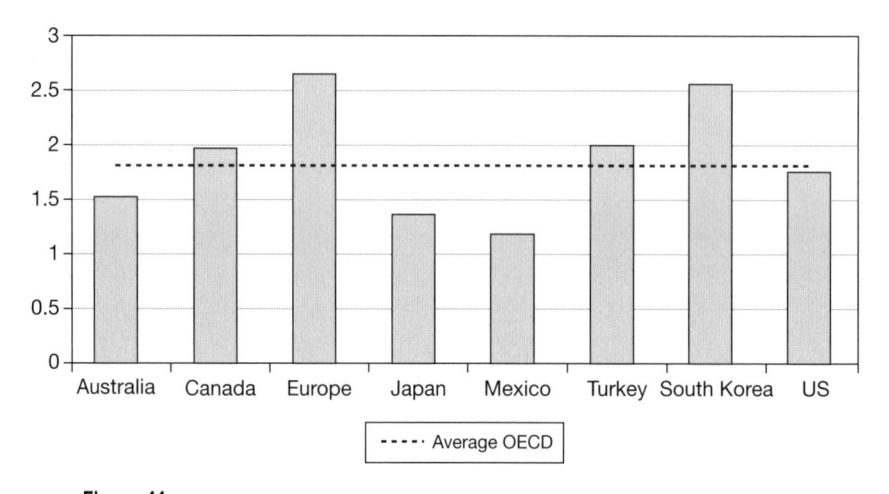

Figure 41

Max month / Min month in major OECD countries.
Source: SG Cross Asset Research, IEA.

If we focus on Europe, the range goes from a maximum in Sweden (4.5) to a minimum in Greece (1.3), in line with the fact that cold weather has an impact on residential demand. And, in 2010, when we saw record low temperatures, this metric hit 5 for Sweden!

SWING PRODUCTION

As gas is used for residential purposes, it leads to huge swing in gas demand depending on the weather. To balance this weather dependant demand, supply has to be flexible.

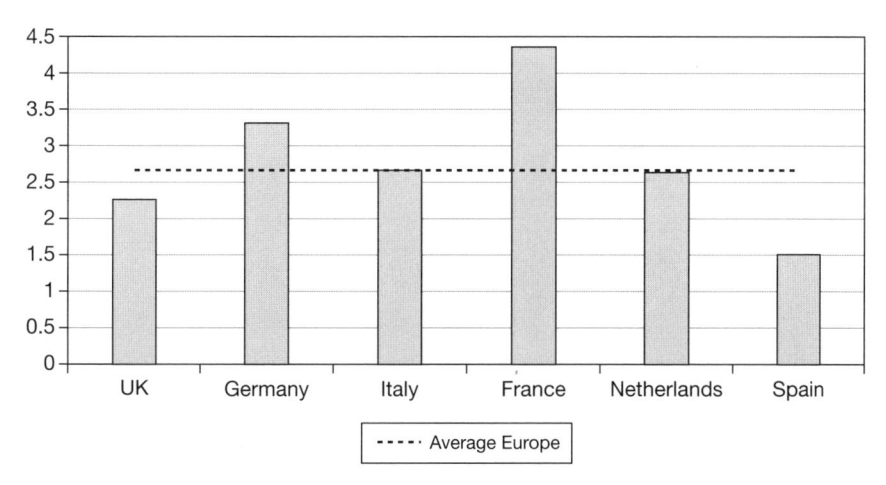

Figure 42

Max month / Min month in major European countries.
Source: SG Cross Asset Research, IEA.

– Three major categories of hydrocarbon fields
In short, hydrocarbon fields can be split into three categories:

- Oil only field.
- Oil and gas field.
- Dry gas field.

In non-OPEC countries:

- All oil only fields are producing at (or near) maximum capacity.

- Oil and gas fields are producing "associated gas", whereby gas has to be separated from oil, but the quantities of gas cannot be tuned to accommodate demand. As oil is being produced at maximum capacity, gas is also flowing at a steady rate (except when maintenance occurs) all year long.

- On the other hand, "dry gas" fields are much more flexible and production there can be "swung" to balance demand changes. This is particularly true for fields next to the big demand areas.

– UK gas production and swing production
As such, the "associated gas" production of the UK North Sea has no seasonal patterns when, on the other hand, "dry gas" production used to be highest in winter and much lower during summer.

Figure 43

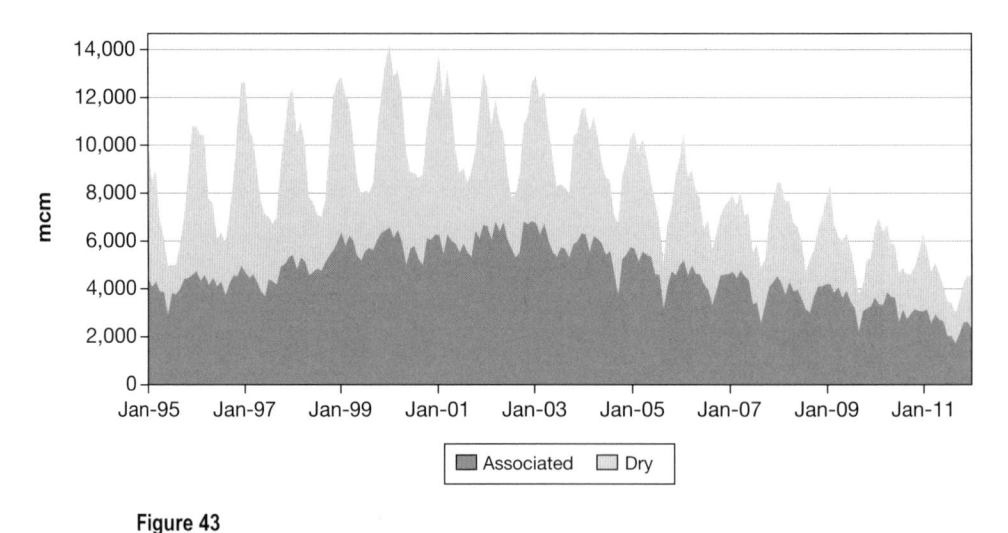

UK associated and dry monthly gas production.
Source: UK Department of Energy & Climate Change.

The decline of UK Continental Shelf (UKCS) means a decline in yearly production figures as well as a decline in swing production. Winter after winter, peak UK production capacity is reduced which forces the UK to become a big net importer in cold months.

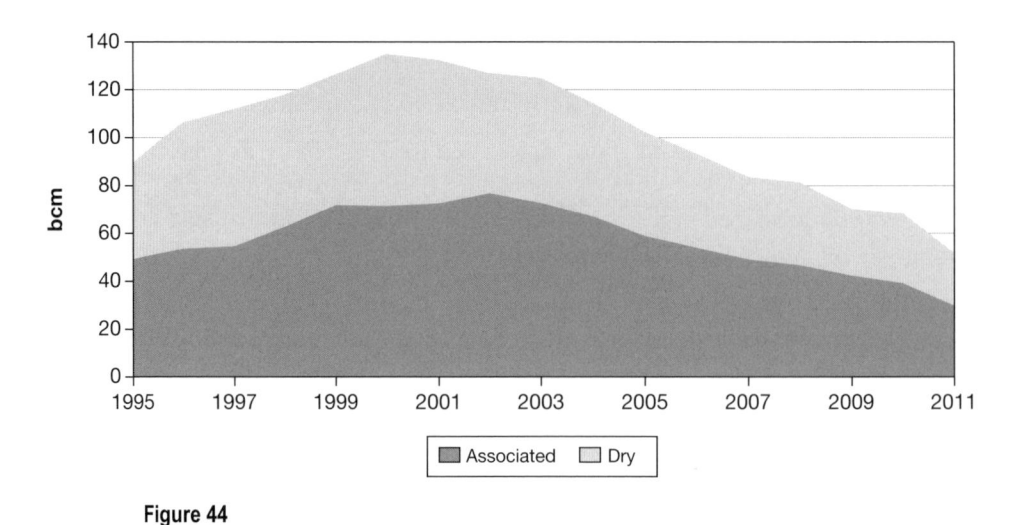

Figure 44

Split of annual UK gas production.
Source: UK Department of Energy & Climate Change.

Gas production peaked in 2000 and then declined by 8.3% on a CAGR in the last 11 years. The production decline has accelerated as it is now 11.1% in the last 5 years.

To define the swing production we have designed a metric with:

- (Max monthly volume – Min monthly volume) / Min monthly volume.

And this over a period of 3 years (as fields in the UK are declining) except for LNG where we only took into account 2011.

This gives a swing factor of 182% for UK production, split into 277% for dry gas field and 142% for associated gas.

– Gazprom: a lot of flexibility... but mainly for Russia!

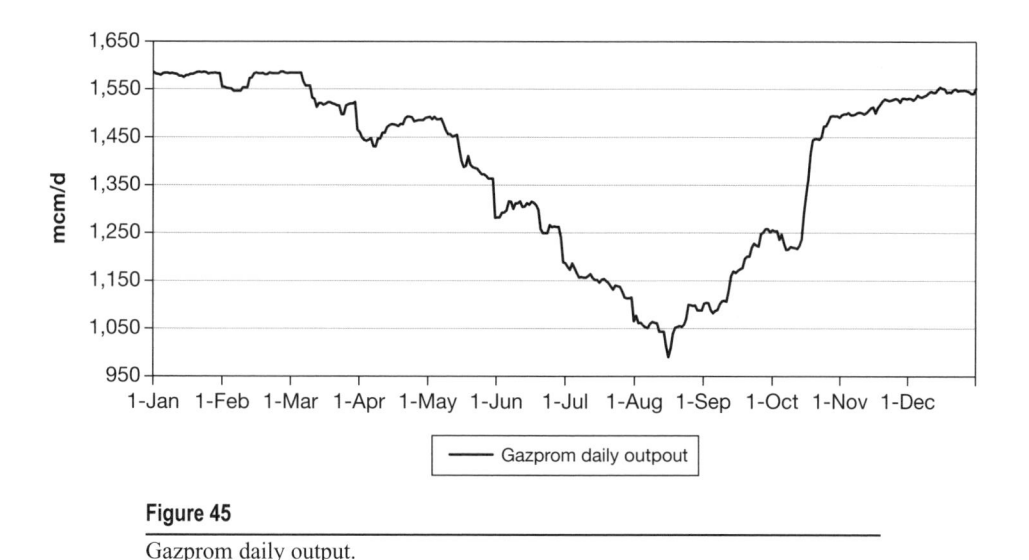

Figure 45

Gazprom daily output.

Gazprom's production is more flexible in Russia than what reaches Europe because, on top of upstream capex, Gazprom also needs to allocate transportation capex, which needs to be recovered by having a high load factor, all year long. As its new production fields (Yamal) are further away, to answer demand seasonality, Gazprom is investing in storage.

Gazprom, even further away from European customers, also swings its supply to Europe by reducing it to as low as 8.3 bcm/month in summer and as high as 16.6 bcm/month in winter (100% swing factor over the last 3 years). This swing allows Gazprom to meet its contractual obligations.

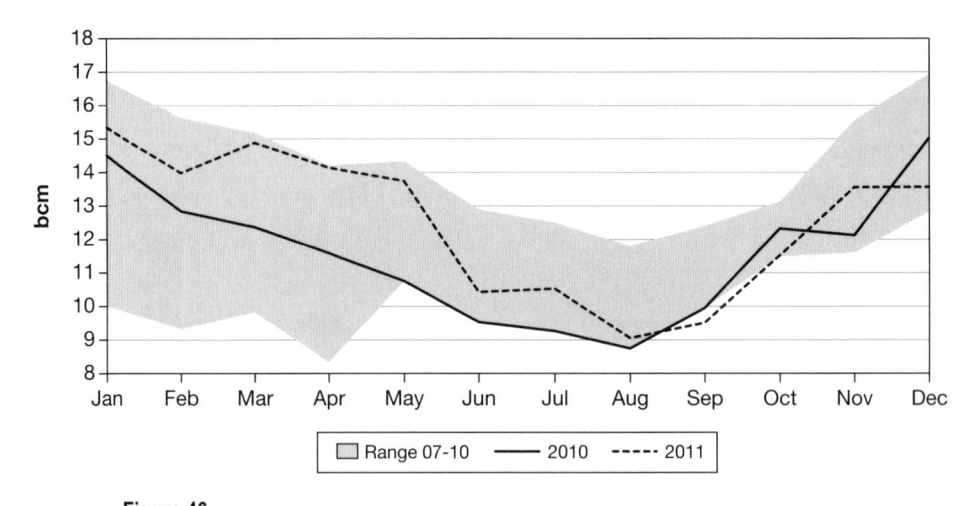

Figure 46

Gazprom monthly exports into Europe.
Source: SG Cross Asset Research.

– Norway production has some flexibility

Norway is slightly closer to European customers but it swings less its production by reducing it to as low as 5.7 bcm/month in summer from as high as 10.9 bcm/month in winter (90% over the last 3 years).

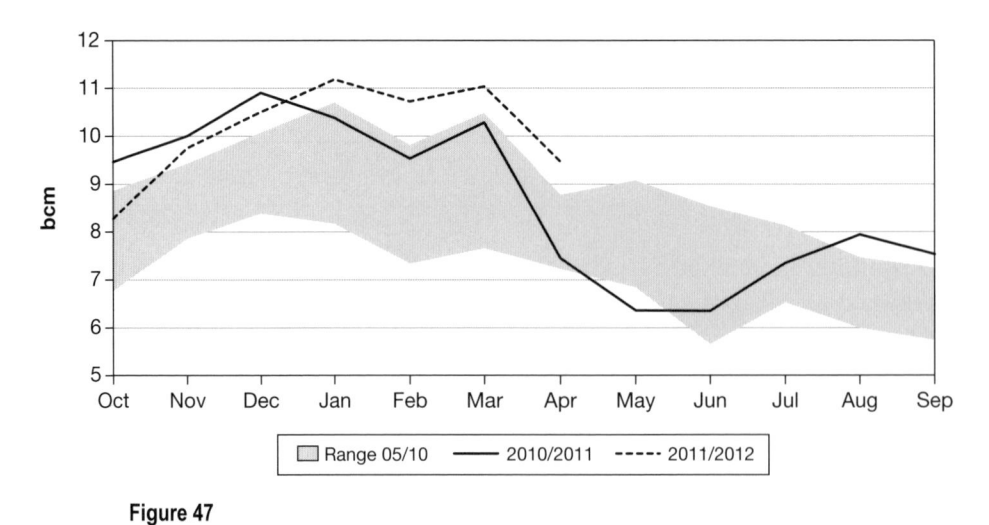

Figure 47

Norway monthly production.
Source: SG Cross Asset Research, Norwegian Petroleum Directorate.

In Norway, production permits are delivered on a "gas year" basis (October to September), hence the above graph.

– LNG: little flexibility on the production side...

Swing production is extremely limited in the LNG supply side. Production is sometime stopped due to war (Libya, from March 2011 to June 2012) or technical problems (Norway in May-June 2011). LNG production can also be negatively impacted when upstream production is reduced (Algeria) or diverted to local market to meet growing domestic demand (Egypt) or affected by political unrest (Nigeria). To compute the swing of LNG we have only taken into account the data of 2011 due to the major growth that was witnessed over the past 3 years.

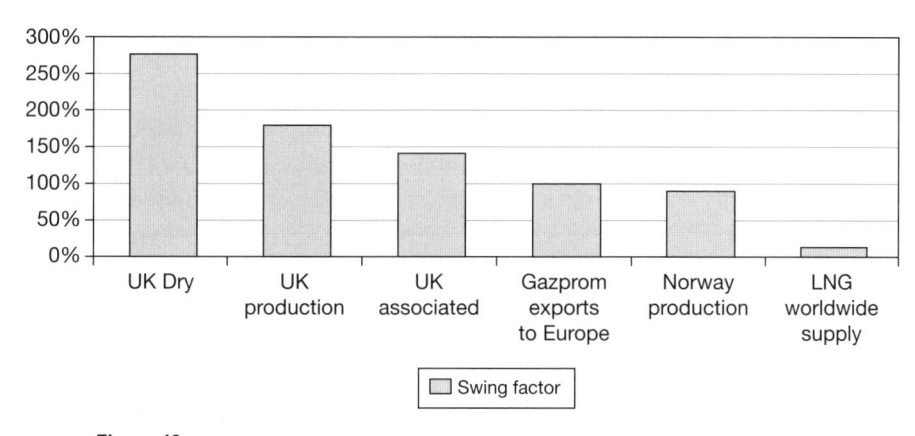

Figure 48

Swing factor.

But this poor swing in LNG supply could still translate in higher swing in receiving countries if they accept to buy LNG on a spot basis when needed.

STORAGE

Long distance pipelines are not designed to carry the importer's peak demand day volume, as that would be too expensive. Instead, the importer relies on other means of meeting peak demand, such as storage, swing contracts with other producers, or LNG imports.

Although underground storage facilities are complicated and costly to construct, they are still the cheapest and, at the same time, safest way of storing gas.

• Three Types of Underground Gas Storage...

Basically there are three ways of storing gas underground – salt caverns and two different storage reservoirs (depleted fields and aquifers). They differ with regard to the reservoir rock and the storage mechanism.

– Salt caverns

Caverns are large natural or man-made underground cavities. The cavities are created by leaching them in rock salt or by using mine workings. The artificially leached salt caverns are particularly important for storing gas underground.

Bore holes are used to inject water into the deeper rock salt layers and to pump the dissolved salt to the surface as brine. The cavern is in most cases cylindrical in shape. Depending on the size, the heights of such caverns vary between 100 and over 500 metres and the volumes of gas stored between 40 and 100 mcm per cavern. The cavities formed resemble underground tanks and the borehole is the only way of injecting gas into or withdrawing gas from them. Compressors are used to inject the gas into the caverns and store it there under pressure. This way, it can be withdrawn quickly at any time for peak shaving. Salt caverns facilities are used for cold days and for trading.

– Reservoirs – depleted fields and aquifers

Storage reservoirs are mainly used for covering the seasonal base load as they often have a large storage volume and, due to the natural flows in the reservoir rock, mostly sandstone, react more slowly to changes in withdrawal rates in the storage well holes. Injection takes place during summer when gas is in excess and withdrawal is used all winter long when supply is short.

A. Depleted oil/gas reservoirs

They are the most common underground storage facilities. These reservoirs are naturally occurring. They have been depleted through earlier production. A good practice is to stop gas production with some gas left in the reservoir as this gas will then become the cushion gas (see below). Otherwise, to convert a depleted field into a storage field, the cushion gas needed has to be reinjected.

B. Aquifers

Those reservoirs are bound partly or completely by water-bearing rocks.

– And two types of above ground storage

To make things more difficult, some countries/companies are adding LNG storage capacity to the overall picture. This storage can either be:

- "LNG peak shaving units" purposely designed for contingency against the risk of emergencies such as system constraints, failures in supply or failures in end user interruption.
- Tanks on a regas terminal.

• ...and two "Kinds" of Gas!

The gas held in a gas storage facility is always divided into **cushion** and **working** gas.

Cushion gas is the volume of gas that is necessary to ensure the minimum storage pressure necessary for optimal gas injection and withdrawal. In caverns, the cushion gas is also necessary to ensure stability. The proportion of cushion gas is about half of the maximum storage volume and remains permanently in the storage facility. In salt caverns, the ratio is about 20%.

Working gas is the gas volume which can be stored or withdrawn at any time in addition to the cushion gas. It is also called "mobile gas".

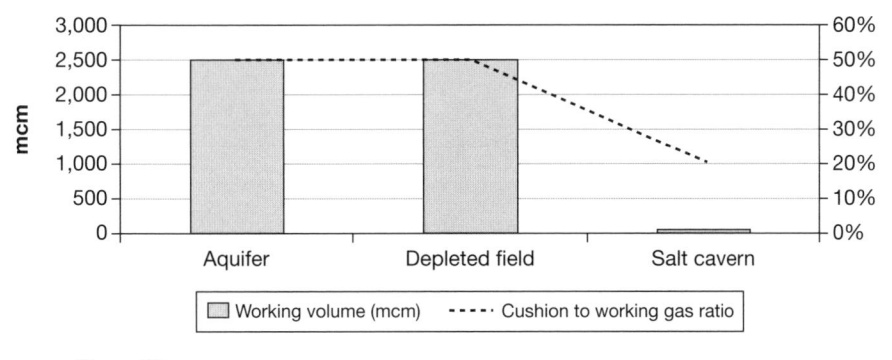

Figure 49

Typical storage figures.
Source: SG Cross Asset Research.

• Worldwide Storage Capacity

With 333 bcm on a worldwide level, storage capacity represents 10% of annual worldwide demand. Storage allows gas to be a complementary fuel to renewables, as gas-fired power plants can be put on-line with very short notice to mitigate a drop in renewable electricity (no wind and/or no sun).

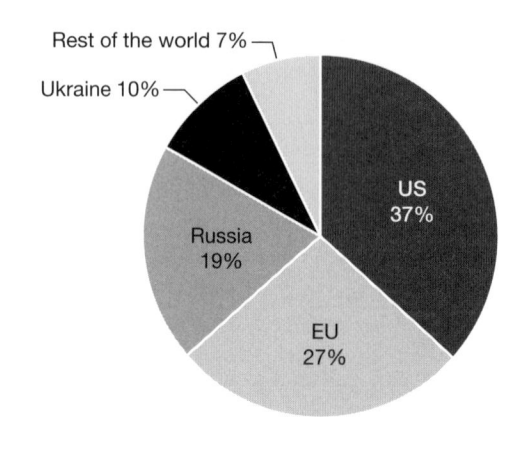

Figure 50

Split of worldwide storage capacity.

– *Storage in the 3 major markets*

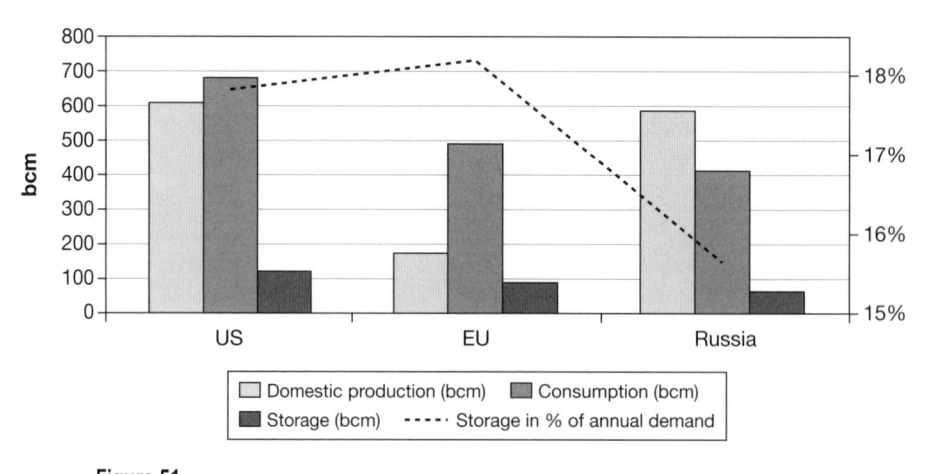

Figure 51

Storage capacity for the 3 major consumers in 2010.
Source: SG Cross Asset Research, BP Statistical Review, GIE, Gazprom.

The three major gas markets experience very different trends:
- The US (22% of worldwide consumption) is becoming self-sufficient but has a lot of storage that can be used for trading purposes.

- EU (16% of worldwide consumption) is the most import-dependant of the three regions and needs storage for seasonal balancing.
- Russia (13% of worldwide consumption) is a gas exporter; thanks to its swing in dry gas fields it needs less storage.

After reaching a low of 16% in 2008, the storage capacity/annual demand metric in EU has moved up to 19% in 2011, thanks to increased capacity and (mainly) reduced demand.

In EU, the total current storage capacity is 90 bcm (and an additional 2 bcm for "LNG peak shaving"). But the split is very different depending on the country considered (history, geology and energy mix). And the demand seasonality is country dependant.

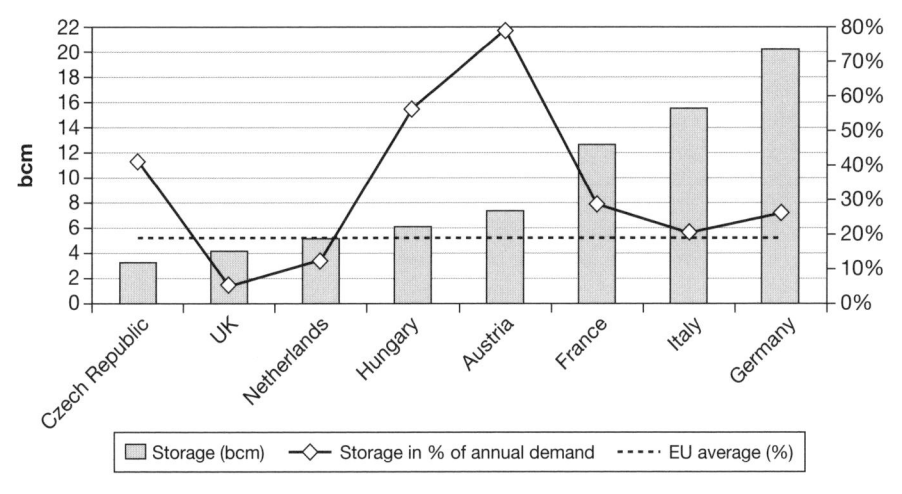

Figure 52

EU countries with major storage capacities in 2011.
Source: SG Cross Asset Research, GIE, Eurogas.

As we've seen that European demand is very seasonal, we can now view the different options to balance this on a summer/winter basis.

According to this measure, Europe needs 71 bcm, with a total capacity of 78 bcm. As we will see, with the following focus, any change in the above equation (change in demand, change in production and /or change in imports) has an impact on storage needs...

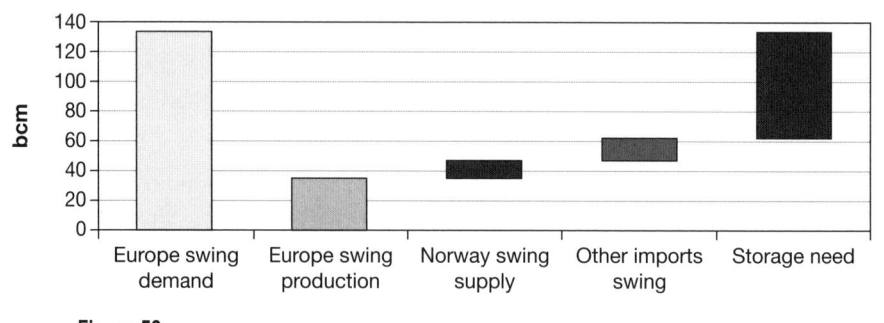

Figure 53

How to balance Europe seasonal gas demand in 2012?

• Focus on UK Infrastructure: Storage vs. Imports

If we look at the winter-summer spread for demand, domestic production and net imports, we see that, back in 2000, this was balanced by storage.

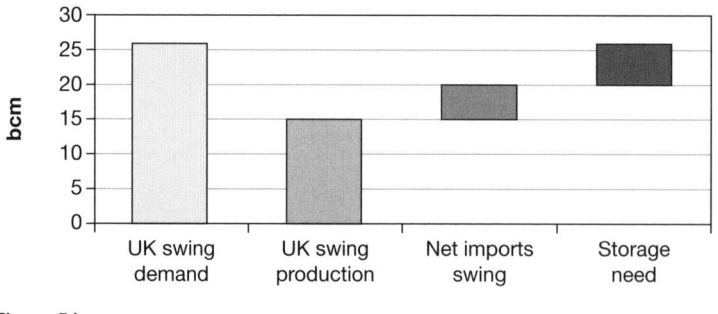

Figure 54

Seasonal balance of the UK market in 2000.
Source: SG Cross Asset Research.

The decline in swing production in the North Sea should have triggered an increase in storage capacity that in reality failed to materialise as companies found it more attractive to invest in LNG regasification terminals with the hope of taking advantage of lucrative arbitrage opportunities, viewing storage as a low return asset.

The rebound in Summer 11 of the production swing is misleading as it is linked to a huge drop in UK production due to operational issues. Without those hiccups, the production level in Summer 11 would have been close to that of Winter 10/11, i.e. with very small production swing…

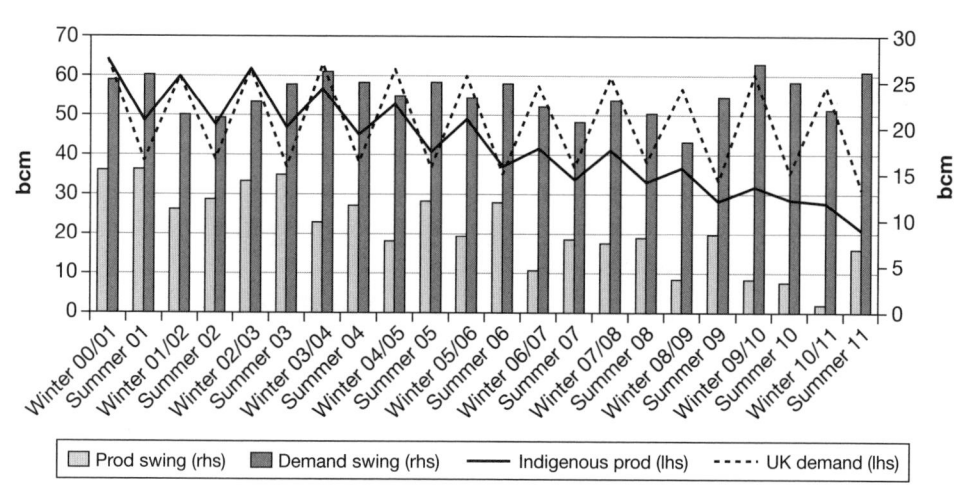

Figure 55

UK seasonal production and demand.
Source: SG Cross Asset Research, IEA.

– But, the market has found another way to deliver this seasonality

Increased imports swings have helped the UK balance seasonal supply. As this was cheaper than investing in new capacity, very little storage has been built in the UK in the last decade even though the UK had already fairly limited storage capacity.

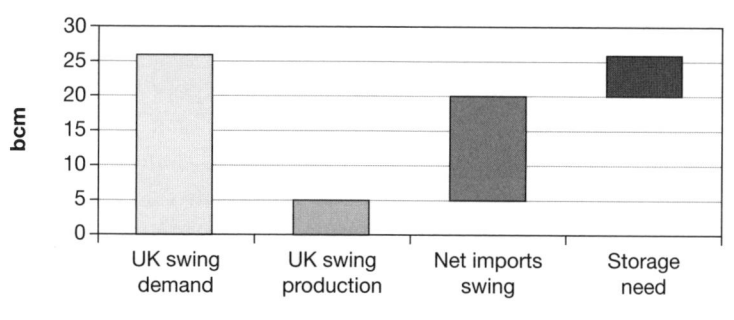

Figure 56

Seasonal balance of the UK market in 2010.
Source: SG Cross Asset Research.

This balancing of the winter-summer demand by other means than storage explains why the winter-summer spread has declined and storage has, in turn, lost value, as we will see later.

• **Russia: Gazprom Monopoly with 65 bcm**

Gazprom has a storage monopoly in Russia and is expanding its capacity both in Russia and in the EU. Since buying Beltransgas, Gazprom also owns and operates all storages in Belarus.

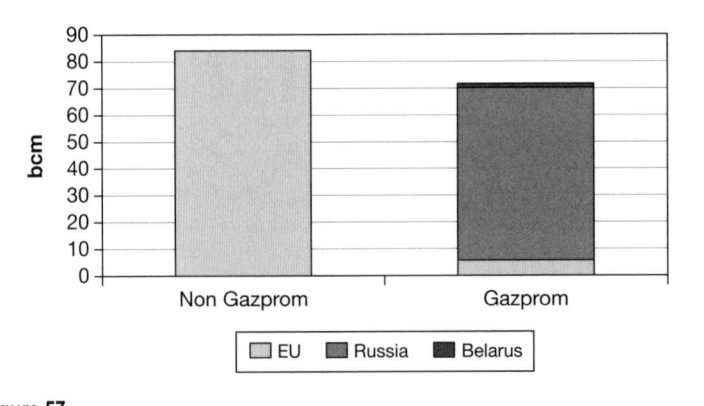

Figure 57

Gazprom total storage worldwide capacity amounts to 71 bcm.

Gazprom has a target to achieve 100 bcm storage capacity in Russia by 2020 to be able to reach a maximum domestic withdrawal capacity of 1 bcm/d.

• **Ukraine: a Special Position in between Russia and the EU**

Ukraine, a major transit country for Russian gas exports into Europe, controls the 4th largest worldwide storage capacity (32 bcm), for a 2010 consumption of 51 bcm. With 62% of storage capacity vs. demand, Ukraine has relatively more storage than both EU (18%) and Russia (16%). Out of the 32 bcm capacity, 27 bcm situated near the European border could be used by European companies to balance supply and demand in winter and help Ukraine remain a dominant gas actor, but Ukraine's poor legislation and high corruption are making this use unlikely.

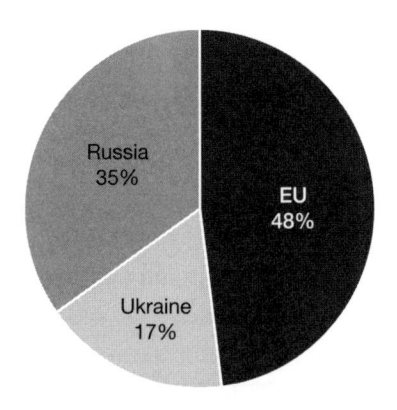

Figure 58

Split of storage capacity between EU, Ukraine and Russia.
Source: GIE.

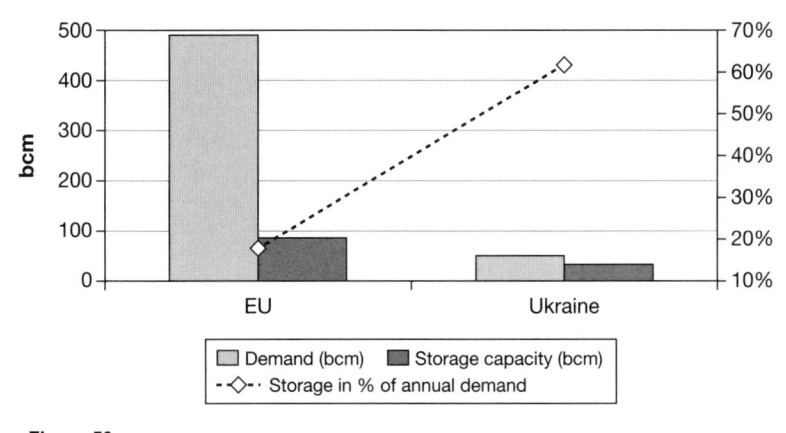

Figure 59

Demand and storage capacity in EU and Ukraine in 2010.
Source: GIE, BP Statistical Review.

SPARE CAPACITY

In oil, OPEC members serve as the "swing" producers in the world market, because only OPEC producers (and in fact mainly Saudi Arabia) possess surplus or "spare" oil production capacity. IOCs companies have no interest in having spare capacity in oil as this entails stranded capex.

For gas, this is different as all producers are normally producing at full capacity in winter and are swinging supply to reduce production in summer, due to the high seasonality of demand. Production maintenance is therefore concentrated in summer, because then capacity reduction has little impact on the supply-demand balance.

– For pipe gas, spare capacity in summer thanks to the swing in production

Assuming Gazprom can reach its maximum daily output all year long, Gazprom would have been able to produce an extra 70 bcm in 2011 (but mainly in summer).

Norway also generated some spare capacity in 2011 by cutting its production in front of a depressed European gas demand. This "value over volume" strategy generated 8 bcm of spare capacity in Norway production in 2011, on top of the 30 bcm swing.

But like all IOCs, Gazprom and Statoil try to produce at maximum capacity in winter.

– For LNG, no spare capacity

On the LNG supply level, relatively poor load factor of 2009 (74%) could be viewed as producers shutting down their trains longer to accommodate the drop in gas demand. Hence in 2009, they were some spare capacity in liquefaction trains. In 2011, with a record 87% load factor, we can assume there was no spare capacity in liquefaction.

Markets, Prices & Costs

The difficulty in the gas business is to plan investment all along the chain in order to meet the requirements of customers all year-long. As it takes decades and billions of dollars to develop major new fields, it is of primary importance for producers to have customers willing to pay for the gas. But it is also important for customers selecting gas as a fuel for their home or industrial process to be sure to get it when needed. We will now look at the pricing issue.

"What is a fair price?" is a question often asked for natural resources as producers and consumers have divergent views. Without elaborating on this philosophical question, we can answer, in short, too low prices stop upstream investments (as seen in the US) and let the demand grow, while too high prices destroy the demand (as seen in Europe), enven through companies have an incentive to make upstream investments.

In a liberalised market, price should reflect cost (at least in the long run). This idea was behind the push for liberalisation, as many producers could compete to produce. But this doesn't apply to natural resources where there is a scarcity rent. For example, with a production cost from less than 10 $/b (in the Middle East) to 60 $/b (for deep offshore or unconventional) for oil and a small transportation and storage cost and with an international oil price at around 110 $/b, the "scarcity rent" amounts from 40 to 100 $/b, making the oil production an extremely profitable business (60% margin on average).

MANY PRICE MECHANISMS...

Gas is mostly traded in the US and the UK. In the US the price discovery is mainly linked to a gas to gas competition. In the UK, it is also a gas to coal competition, as gas and coal can both be used for power generation. In continental Europe and Asia gas is mainly priced under oil-linked formulas with a 3-9 month delay (hence following the Brent price with a lag).

Figure 60

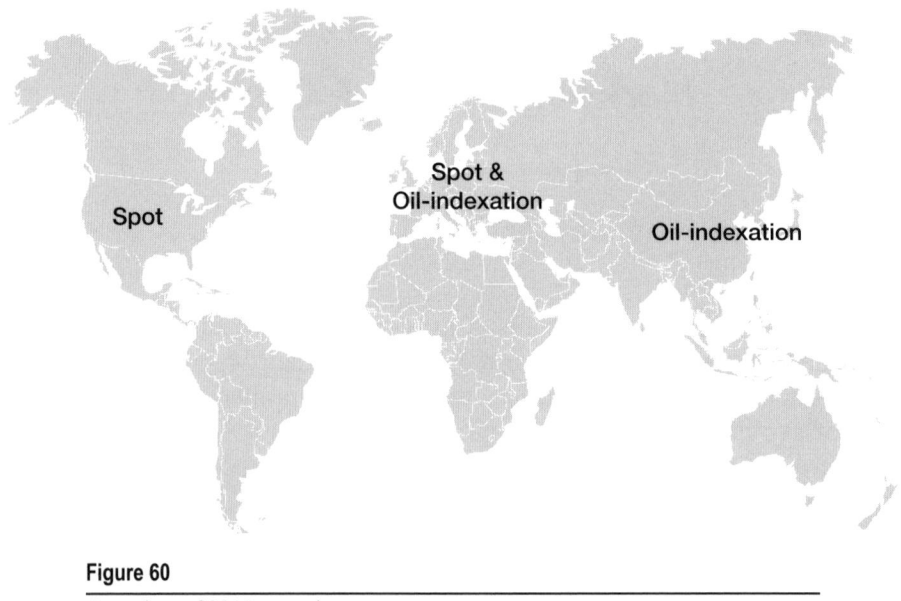

Overview of 2012 gas prices.
Source: SG Cross Asset Research.

...& MANY UNITS

Prices are in $/MBtu in the US and Asia, in p/th in the UK, in $/1,000 cm for Russian gas sold in continental Europe (linked to oil). To be able to compare those prices we have to use the same unit. The difference between 2 prices (in the same unit) is called a spread.

• Spot

As under market rules, prices should reflect costs. When liberalisation of the gas sector started in the US in 1978, the objective was to move to a full market mechanism with spot, futures and forward prices, to allow counterparties to exchange risks. In liberalised countries, the price discovery mechanism is done by market places. It is important to underline that the only signal given by a market is the price.

The reference US price is called Henry Hub (HH), named after the place in Louisiana where major pipelines connect. Gas prices in the US are now linked to the physical balance of supply and demand. And as supply is growing faster (thanks to shale production) than demand, prices are depressed, pushing companies to divert capex away from gas production...

The UK followed with the privatisation of British Gas in 1986 and the establishment of the National Balancing Point (NBP) in 1996. The NBP price is higher as UK, so far, has not managed to unlock its shale reserves.

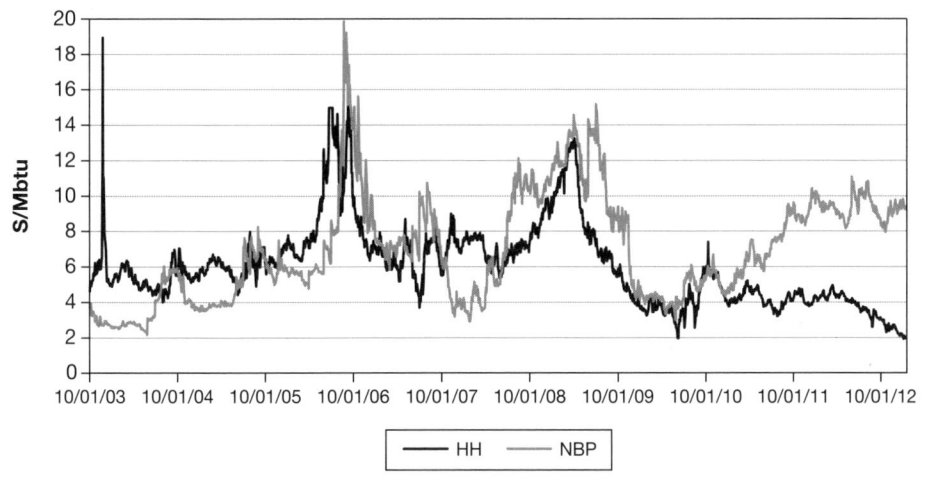

Figure 61

NBP vs. HH.
Source: SG Cross Asset Research, Datastream.

• Oil-indexation

European natural gas development accelerated in the late 1950s with the development of the giant Groningen field in the Netherlands. Before 1960 there was very little international trade in gas in Europe, but in the early 1960s the Dutch began to negotiate with Germany, Belgium, and France for the export of substantial volumes of gas by pipeline. As end-user gas prices across much of Europe were state controlled, and free markets did not exist, the new Dutch concept of gas pricing was established in 1962. In order to generate maximum revenue for the state, the "netback" principle was introduced as the basis for gas marketing, as opposed to the prevailing principle of cost-plus. The distinction between these two approaches is that the cost-plus methodology is additive, but the netback value approach is subtractive. Cost-plus pricing starts with the production cost, and adds transportation services, overheads, and profit margin, to arrive at the sales price. Netback pricing begins with the "market value" of gas in inter-fuel competition (in each market sector) and deducts the costs of transport services and overheads to arrive at the "netback value" at the point of production. If gas was to capture market share from other fuels, then companies would need to incentivize customers to invest in gas equipment through competitive pricing. The principal competing fuel in the domestic sector was gasoil, the heating fuel of choice. Commercial consumers also used much gasoil, but larger consumers frequently used the much cheaper heavy fuel oil as the primary source of heat for both industrial processes and space heating. The "netback prices" are different depending on the market reached. The producers get the netback prices that must cover for the production costs and the margins. This principle did, as expected by the Dutch, generate a rent that was much higher than a simple economic margin.

Long-term contracts are internationally binding agreements and the gas under such contracts will therefore always flow except in cases of "force majeure". In Europe the first long term contract, was signed in 1962 to monetise gas from the Dutch Groningen field. As at that time, they were no gas market, a formula was devised to price the gas. The idea was to push customers away from oil products into gas and therefore the gas price was linked to the price of petroleum products with a small discount. To avoid volatility, the price was set on a quarterly basis taking into account averages of the past 6 to 9 months of the competitive oil products. This formula called "oil-indexation" gives by construction visibility but doesn't take into account the balance between supply and demand of gas.

This oil-indexation was also applied to long term LNG contracts. Each LNG contract is specific. But the general price formula was based on an S-curve where, between a floor and a ceiling, gas prices were correlated (on a fuel switching basis) to oil and oil products. Outside the correlation band, the gas price was fixed. This means producers should still receive a minimum amount of cash if oil prices go down too much and buyers cap their bill if oil prices go up too much. Some old LNG contracts have a ceiling as low as 25 $/b for oil!

Gazprom (for pipe gas) and Qatar (for LNG) are still pushing for oil-indexation for their long term gas contracts.

This graph shows the consequences of the US shale gas revolution: US prices are much lower than anywhere else in the world. With equivalent prices below 15 $/boe, gas is very cheap in the US. We will later discuss the future of oil-indexed long term contracts.

TERM MARKETS

Term markets give a price to the commodity for future delivery:
- The forward market is the over-the-counter financial market. Forward contracts are personalized between parties (i.e., delivery time and amount are determined between seller and customer).

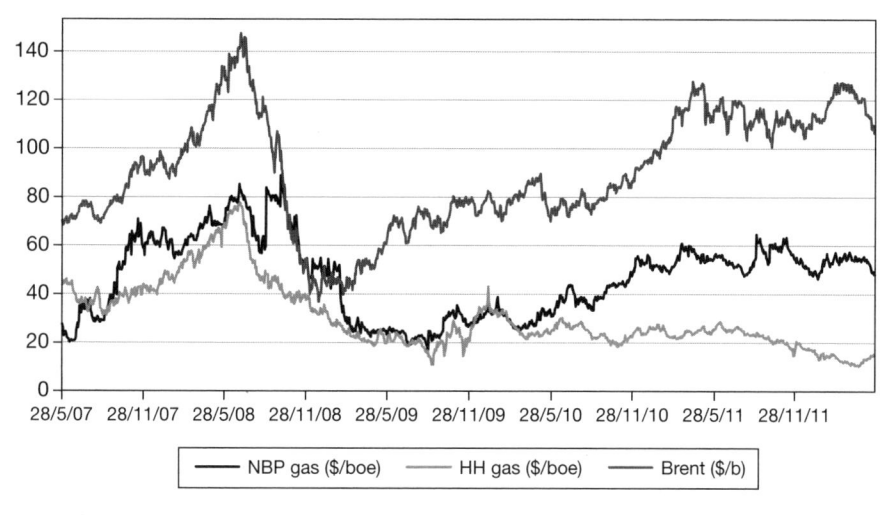

Figure 62

NBP gas, HH gas and Brent oil prices.
Source: SG Cross Asset Research / Datastream.

- Futures market is a central financial exchange where standardized futures contracts (with delivery set at a specified time in the future) are traded. The contracts traded on futures exchanges are always standardized. To make sure liquidity is high, there is only a limited number of standardized contracts.

The New York Mercantile Exchange (NYMEX) is a commodity futures exchange for standardized futures contracts:

- For oil (delivery of West Texas Intermediate to Cushing, Oklahoma), the contracts also serve as one of the key international pricing benchmarks.

- For gas (delivery at the Henry Hub in Louisiana), the contracts are widely used as a US benchmark price.

In Europe, the IntercontinentalExchange (ICE) is the market place for Brent crude oil and UK NPB natural gas, just to name the most well-known contracts.

Markets offer opportunities for risk management of the dynamic (volatile) pricing of gas: companies wishing to hedge their risks will find counterparties (other companies that have a reverse risk or speculators wishing to warehouse the risk). A liquid market will need as many buyers and sellers as possible, hence the need for standard contracts.

Contrary to common belief, term markets show little predictability. The price of a future delivery is no guarantee that the commodity will achieve this specific price on this specific month. It is the average level all market participants are comfortable with, at the time of the transaction.

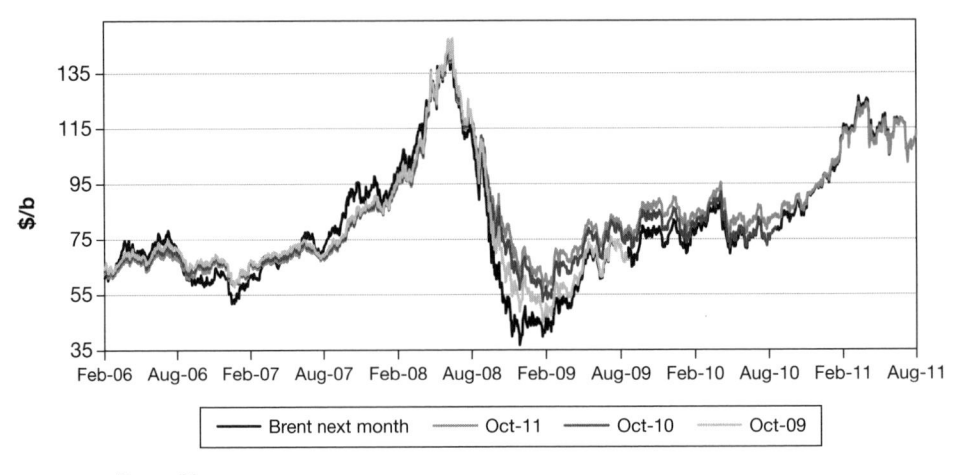

Figure 63

Brent vs. specific deliveries.
Source: SG Cross Asset Research, Reuters.

As we see, the future price for delivery of Brent in October 2011 moved between 61 and 144 $/b during the time this contract was traded…

Figure 64

NBP vs. specific deliveries.
Source: SG Cross Asset Research, Reuters.

And for gas, the future market has rarely been a reliable indicator of actual outcome. Again term markets are there more for risk management than to predict future prices! As we will see later, gas markets can provide a winter-summer spread, reallocate supply or select the most profitable fuel to generate electricity.

COSTS

• Cost Structure in Europe

Now that we have discussed prices, we need to look at the cost structure in this industry.

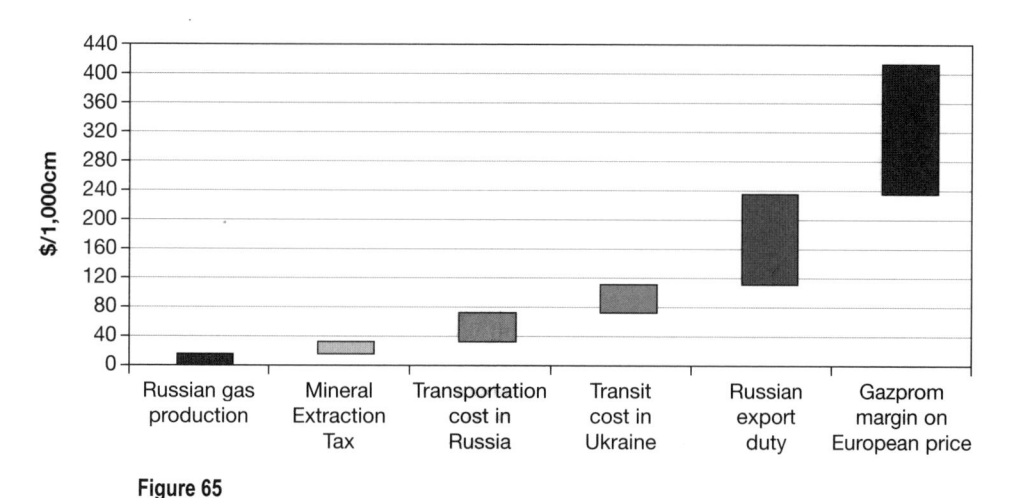

Figure 65

Cost split of Russian gas for Europe.

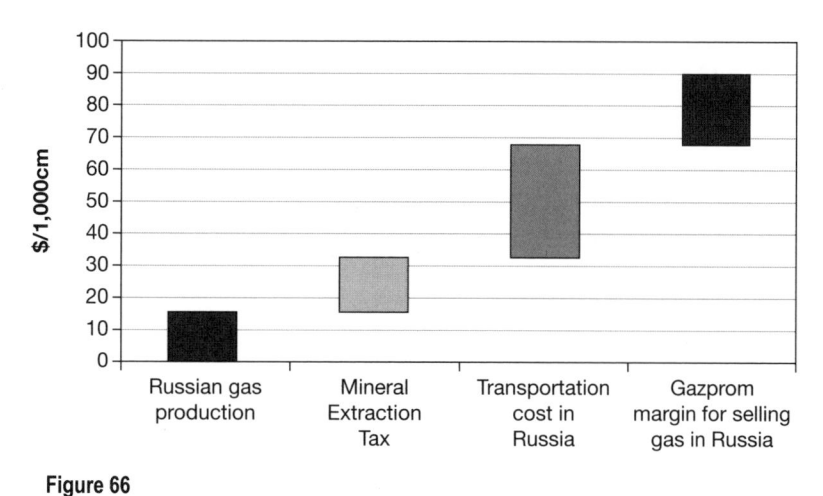

Figure 66

Cost split of Russian gas for domestic market.

With a cost of production for Russian gas at 16 $/1,000 cm, Gazprom sells its gas in a price range from 90 $/1,000 cm in Russia (where prices are regulated for domestic customers) up to 415 $/1,000 cm in Europe (and 320 $/1,000 cm in Former Soviet Union). If we take into account the flat mineral extraction tax (16 $/1,000 cm), the cost of transport and the export duty tax (30% of final price), we end up with a (pre-tax) margin moving from 22 (in Russia) up to 179 $/1,000 cm for Europe… leading to a margin moving from 33 to 76%!

– *Transit risks: dispute over $18.5bn!*

If we take an hypothesis where Gazprom can't transit the gas in Ukraine and has to sell it with 0 profit on the Russian-Ukrainian border, we then have a margin of 272 $/1,000 cm for Ukraine.

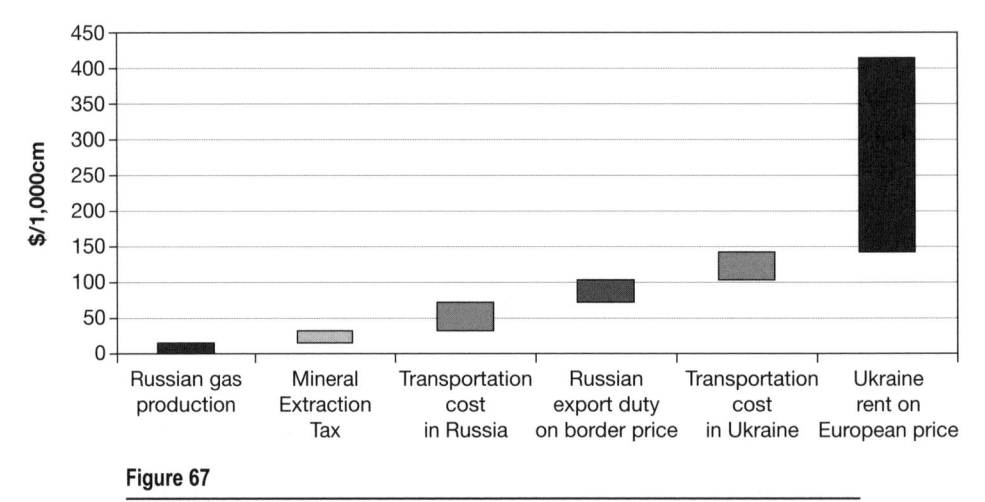

Figure 67

$18.5bn if Ukraine takes all the rent.

For 68 bcm of gas in transit, this equals $18.5bn per year rent on top of the $2.7bn cost for transportation (that we assumed equal to the transit tariff). This explains why Gasprom is trying to use alternative routes. As producers don't want to share their rent, transit risk can only be mitigated by avoiding transit!

• **Cost Structure in the US**

With prices in the US at 2 $/MBtu, the question is what is the cost of extracting unconventional gas. If markets were always perfect, the simple answer would be 2 $/MBtu for the marginal producer (the most expensive one) and lower for all the others. In free markets, price should reflect cost, at least in the long run. But not always in the short term as:

- US producers have hedged some of their future gas production at higher prices, making it still profitable for them to produce.

- In the US the owner of property rights can severe mineral rights from the surface rights. Therefore the subject of the oil and gas lease can be limited to the minerals. The surface interest is what remains of the bundle of rights of land ownership after mineral interest has been severed. The surface interest generally gets a share of production or the value of proceeds of production, free of costs of production when and if there is production on the property. Therefore to avoid company keeping acreages without producing, leases between land owners and companies become void if a company doesn't drill a minimum amount of wells in a specific timeframe or the lessee has to pay the land owner for deferring commencement of drilling operations. Those specific contracts force, lessees to drill even if prices are very low, increasing production and pushing prices further down…

- In joint venture, the new partner has to pay an upfront payment to the US company that has aggregated leases. The new partner is required to pay a substantial portion of future drilling costs. In general, these are time limited commitments. So, the drilling money gets spent irrespective of the market price. In short, new partners pay for the drilling and the JV gets the gas…

Another difficulty in estimating the cost of unconventional gas in the US is the fact that companies are moving away from shale gas plays to shale oil plays where oil makes the investment profitable and gas, that still accounts for 50% of the output could be viewed as a by-product (see below).

Those "imperfections" are blurring the price signal. We estimate the cost of US shale gas production to be above the actual HH price, at around 4 $/MBtu. This means that, with time, when those imperfections will ease, the HH price should move slightly above the marginal cost of production. But as soon as price moves up, producers will re-invest in shale plays capping the price not too far away from this marginal cost.

• Cost of associated Gas

An additional layer of complexity is the fact that gas can either be produced from a dry gas field or as associated gas to oil. If it is possible to estimate the cost of dry gas, the cost of a by-product of oil production (associated gas) can vary according to the hypothesis taken. With oil above 100 $/b some fields are so profitable that associated gas can be taken for free! This is the case in shale oil plays (and in North Dakota, the associated gas is flared due to the lack of infrastructure being able to ship it to far away customers). This explains why gas prices can reach lows of 2 $/MBtu in the US. But this could also be applied in Qatar where the distillates produced are making the whole upstream/liquefaction investment profitable. When oil prices are high, Qatar could give away its LNG for free, but that's not Qatar strategy that is seeking gas prices close to crude oil parity!

EUROPE: FROM OIL-INDEXATION TO SPOT MARKETS

• UK: the Oldest liberalised Market in Europe

UK became the first market in Europe to open up in 1986. Today, in Britain gas anywhere in the country's transmission system counts as being at a single point. The NBP, created in 1996, has boosted trade and is the most liquid gas trading point in Europe. NBP serves as the major European price index, but Norway that accounted, for 28% of the UK total consumption in 2010, has an increasing market power that could be detrimental to the NBP going forward. Norway can, due to the flexibility of its export network, choose to arbitrage UK against Continental Europe or reduce production.

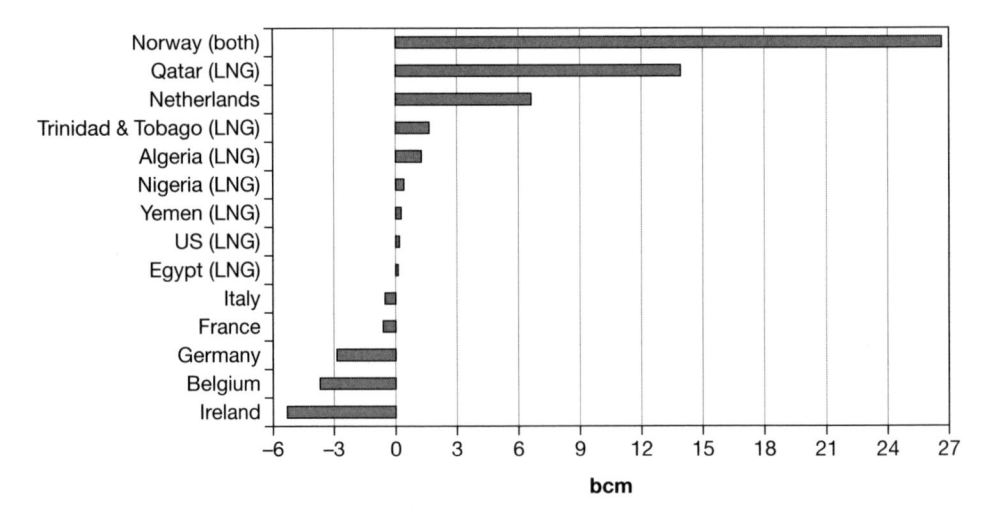

Figure 68

Gas trade movements in and out of the UK in 2010.
Source: BP Statistical Review.

UK is still the largest EU market with 18% of consumption, but in Continental Europe the main buyers are Germany, Italy, France and the Netherlands.

The liberalisation process has been very lengthy in Europe. It started back in 1998, with directive 98/30/EC concerning common rules for the internal market in

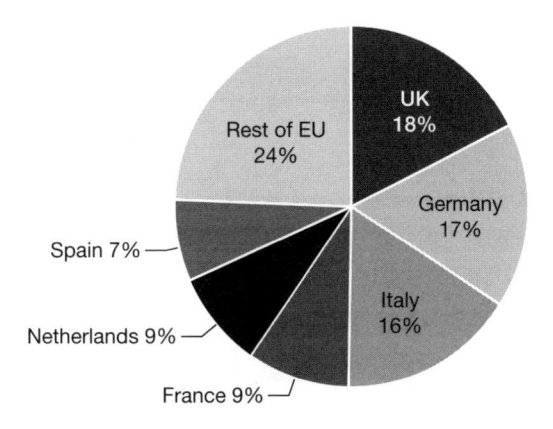

Figure 69

Split of EU 2011 consumption.
Source: Eurogas.

natural gas. The full opening-up of national gas markets is set out in directive 2003/55/EC. In practice, industrial clients and domestic customers have the freedom to choose their gas supplier since 2004 and 2007 respectively. As the internal market in natural gas suffered from a lack of transparency which impedes its proper functioning, the EU considered it necessary to redefine the rules and measures in directive 2009/73/EC (third energy package) in order to guarantee fair competition and appropriate consumer protection...

Continental Europe did resist liberalisation and until 2005, most Continental European buyers were happy to resign full oil-indexed long term contracts (up to 30 years!) with Gazprom. Much lower US gas prices and lower UK gas prices are now putting pressure on those buyers to renegotiate those long term contracts to include spot indexation. So Europe is in the process of moving away from oil-indexation to spot pricing, but this is a lengthy and painful process. In Europe, gas is therefore now sold under both oil-indexation long-term contracts and on a spot basis.

According to our estimates, in 2011, 58% of the gas sold in Europe was under an oil-linked formula. Since the 2008 crisis, this ratio remained unchanged at 58%: successful renegotiations in long term contracts introduced some spot indexation, on one side but, on the other side, due to demand destruction, buyers had to reduce their spot purchases and the increase of Qatari LNG in Continental Europe during this period was mostly oil-linked. But the renegotiations and arbitration cases could reduce oil-indexation to less than 50% of the total, before 2014e. We believe this could be a tipping point as under 50% the system could be unstable.

68 *After the US Shale Gas Revolution*

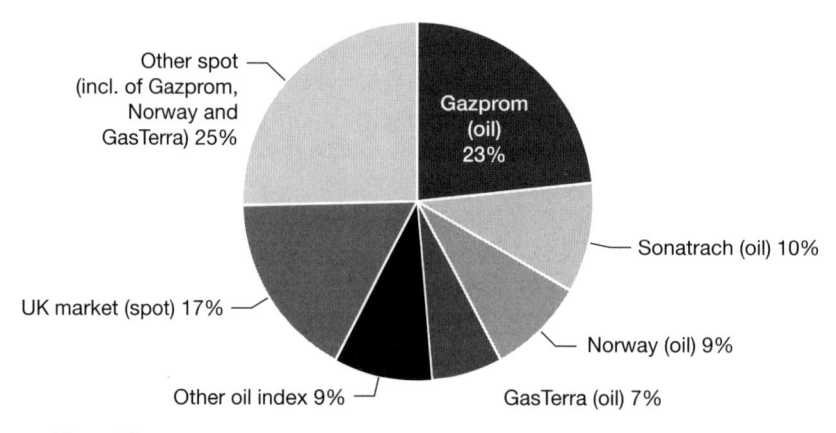

Figure 70

Europe gas supply: 58% oil-linked in 2011.
Source: SG Cross Asset Research.

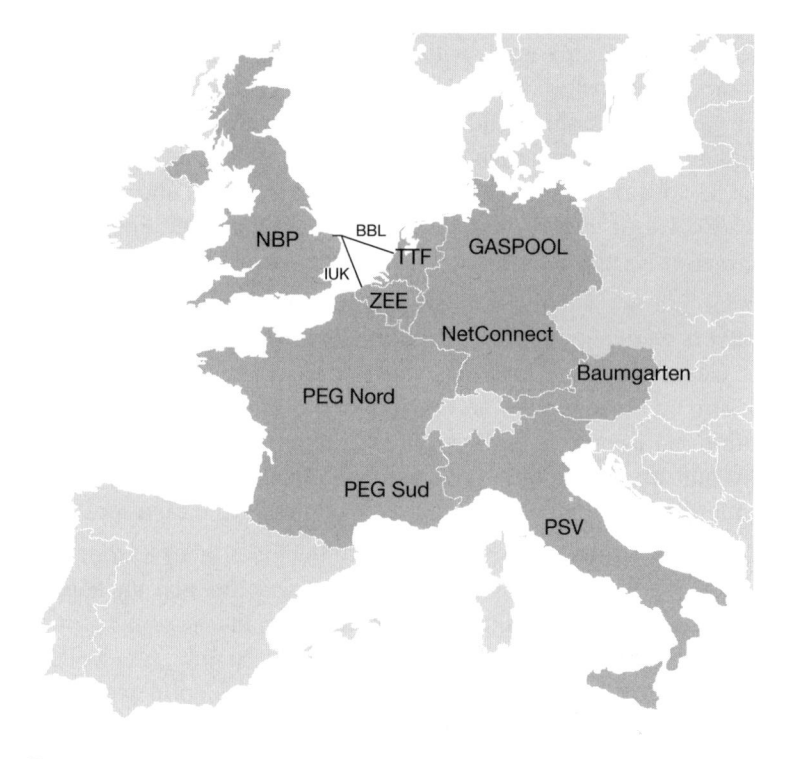

Figure 71

Major hubs in Europe.
Source: SG Cross Asset Research.

– Interconnector: the only two-way pipeline to balance those 2 pricing mechanisms

In 1998, the Interconnector became the first ever bidirectional pipe. It can balance UK and Continental European markets; it is the only pipe that doesn't link a production area to consumers. The differences in pricing between the UK (mainly spot) and Belgium (still mainly oil-linked) set the flow direction: traders move gas from the cheapest to the most expensive place!

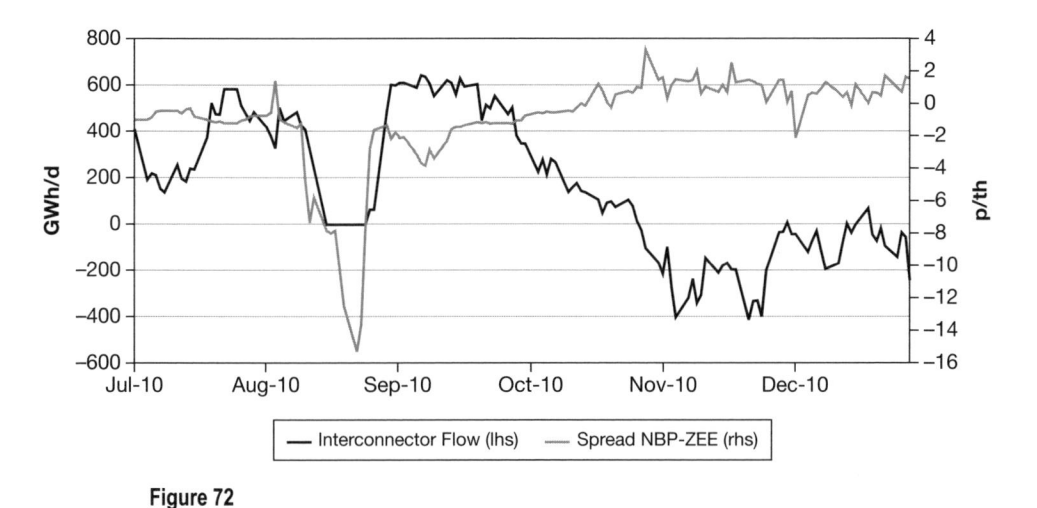

Figure 72

How the Interconnector balances UK and Continental markets.
Source: SG Cross Asset Research – Interconnector UK.

The Interconnector exported the spot market principle to Continental Europe. But it also links UK spot prices to the Continental oil-indexation principle…

Thanks to the Interconnector, the NBP price is highly correlated to the Zeebrugge (ZEE) price. As long as the Interconnector is operational (and not full), the spread between NBP and ZEE is, at maximum, the cost of transporting the gas (less than 3 p/th). But when the Interconnector is not available (for example during the planned maintenance period at the end of September 2010, represented on this graph), then prices are diverging as traders have no way to push the gas to the most expensive market. In theory, if the Interconnector was to operate at full capacity, prices could also diverge if the capacity was not enough to take all nominations from traders. So far, this has never happened.

• **Future of European Hub**

ZEE is also highly correlated to most continental prices (except Italy) thanks to a European gas grid that is getting more flexible. After the 2009 crisis, the European

Commission pushed for reverse flow capacity to be implemented between all major markets, to increase security of supply and to allow more trading in Europe. The more European hubs become interconnected, the fewer hubs we should need for price discovery...

As Europe could follow the US route, it is interesting to see how hubs are competing. To allow the price discovery mechanism to work properly, markets require liquidity, transparency and regulation.

Table 3 European hubs competition

	UK	Belgium	Netherlands	Italy	France	Austria	Germany	
Hub	NBP	ZEE	TTF	PSV	PEG Nord	Baumgarten	GASPOOL	NetConnect
Exchange	ICE, APX-ENDEX			GME	Powernext	CEGH	EEX, ICE	
Storage	---	---	+	++	++	+++	++	
Virtual storage	Yes	Yes	Yes		Yes			
Churn ratio	++	-	--	--	--	--	--	--
Liberalised market	++	+	+	+	---		-	
Transparency	+++	--	++	++	++	--	---	
Trading in €	No	No	Yes	Yes	Yes	Yes	Yes	
Access to LNG	+++	+++	++	++	+	Impossible	No	
Political backing		---		+	---	++	++	

– *Churn ratio*

The churn ratio measures the liquidity on a hub. It is the amount of time a molecule is exchanged on the hub before being burned (consumed).

In Europe with a churn ratio of 12, only the UK NBP meets the minimum liquidity requirement. It is interesting to note that if all trading in Europe was concentrated on one hub, this hub would have a churn ratio of 33, in line with the US HH (30). This shows that concentration in Europe should continue for the benefit of one hub that would then become very liquid. The European target of completing the integra-

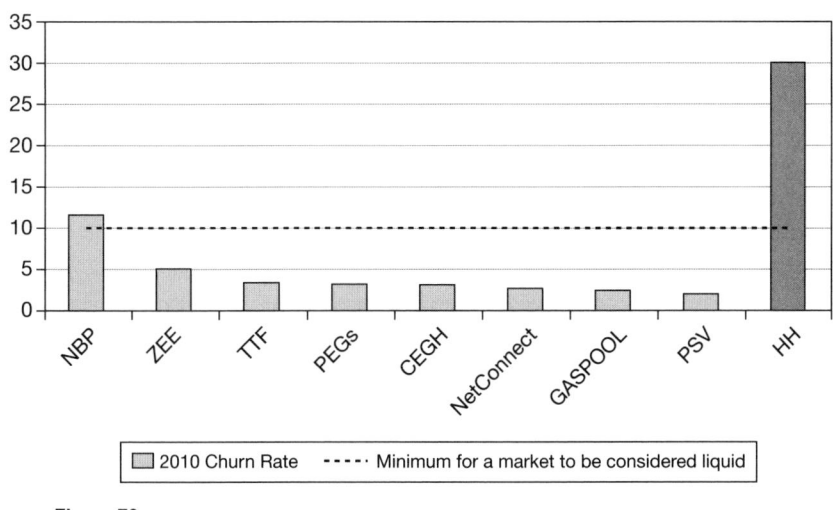

Figure 73

Churn ratio on different gas hubs.
Source: SG Cross Asset Research, IEA.

tion and liberalisation of the internal market by 2014 to allow member states to share and trade energy in a more flexible way could boost the outcome of a major European hub. This would then allow this European gas index to be viewed by the financial markets as a commodity asset as liquid (and therefore as tradable) as the HH.

– Renewed hub competition…

The new Gate regas terminal in the Netherlands is important not only because it provides more capacity but it also allows the Dutch hub Title Transfer Facility (TTF) to access international gas prices thanks to LNG. TTF is therefore in a better position to compete versus the actual main hub, the NBP in the UK. With access to LNG and more storage capacity due to the Bergermeer plant that could be operational in 2013e, TTF could become a major competitor to NBP not only as the number one trading place on the European continent but perhaps as the number one place in Europe, in the long run.

Italy's current liberalization process includes the ownership unbundling of gas network operator Snam. This could free up Snam to develop Italy as a gas transport and trading hub. But even if Italy is well positioned (with Algerian, Libyan, Norwegian and Russian pipe gas and 2 LNG regasification plants), this process is too late to materialise. The final race is between NBP and TTF…

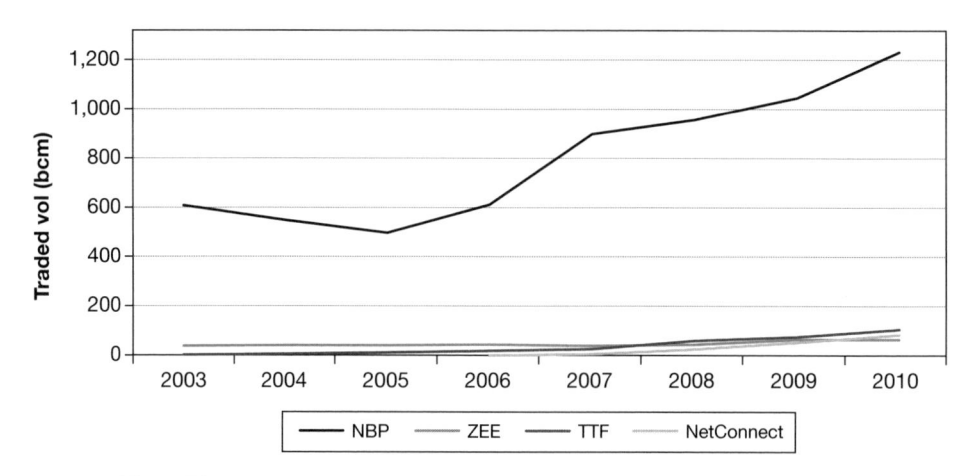

Figure 74

NBP way ahead of competition...
Source: SG Cross Asset Research, IEA.

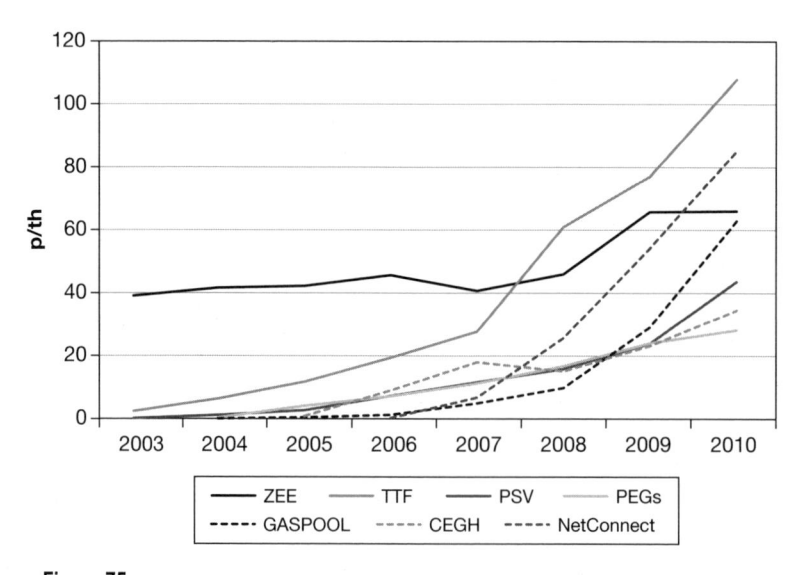

Figure 75

... but TTF is gaining momentum!
Source: SG Cross Asset Research, IEA.

– *...but sill not enough producers!*

Relying on hubs for price discovery makes sense in the US where numerous producers are in competition. This is not the case in EU where the main external sources of supply in 2011 were: Russia (24%), Norway (19%), Algeria (9%) and Qatar (7%), giving those four countries and their state-owned company c.50% of the market. Europe has little choices if it doesn't want to allow domestic shale gas production. Long term contracts (10 years, hub index and no flexibility) could still be used by the gas industry but will not look the same as formerly (30 years, oil-linked and some flexibility).

MARKETS CAN MITIGATE WINTER-SUMMER SPREADS

Before liberalisation, gas storage was built to balance supply and demand (on a seasonal basis), as well as for broader security of supply concerns. With liberalisation, companies started to use storage in a more flexible way as they were no longer obliged to prioritise security of supply.

The industry has to balance seasonal demand with adequate supply. As we've seen this is mainly done thanks to storage. In a liberalised market, the spread between winter and summer prices signals the need for (and use of) storage. The wider the spread, the more valuable the storage, hence the incentive for market participants to invest in new storage capacity. Today's low winter-summer spreads are a market reflection that storage capacity is adequate.

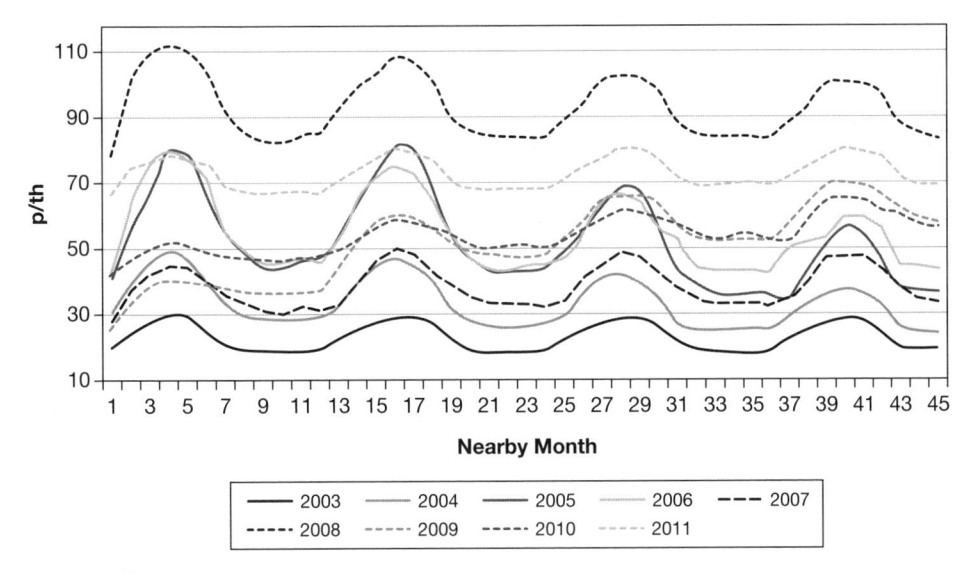

Figure 76

Forward NBP curve on 1st September of each year between 2003 and 2011.
Source: SG Cross Asset Research.

If we compute the winter-summer spread for each year for the next 4 years, this paints a picture where storage has lost value since the 2005 peak in absolute terms (p/th) but also as a percentage of the summer price (%).

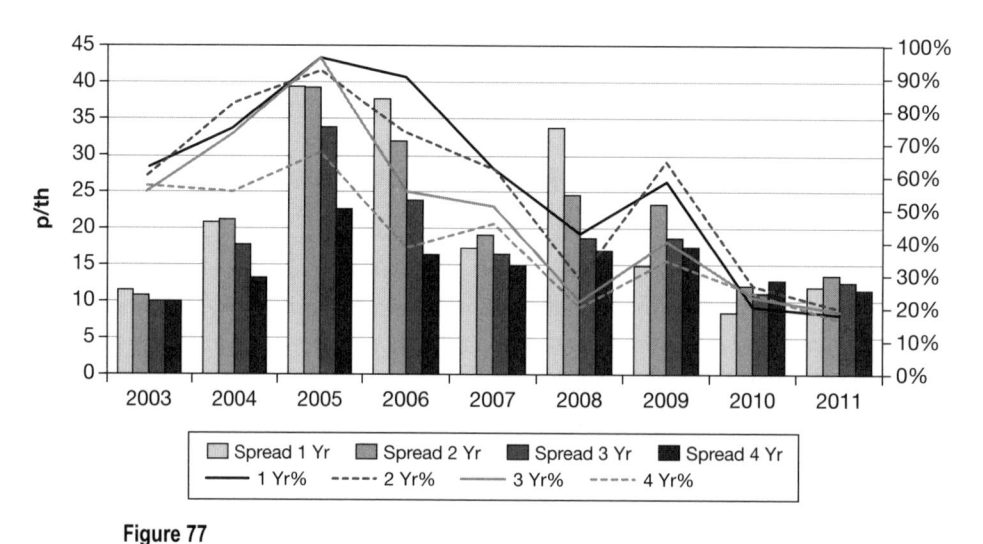

Figure 77

Winter-Summer spread on 1st September of each year between 2003 and 2011.
Source: SG Cross Asset Research.

This drop in storage valuation is in line with what we've seen earlier as storage has been in competition with increased import infrastructure that allowed the UK to balance supply and demand by swinging more imports.

This doesn't reflect any security of supply risk as market can't value low probability/high risk events.

MARKETS CAN REALLOCATE SUPPLY

• LNG Arbitrages

Since the beginning, in 1963, long-term contracts (20-year) have been the backbone of this industry. The entire LNG chain was built at the same time: a liquefaction plant in the producing country, ships in yards and a regasification plant in the consuming country. Arbitrages were previously infrequent. At the turn of the millennium, some companies started to implement a more lucrative business model by redirecting more and more cargoes. This business also helped rerouting cargoes away of the US (since 2010 and the shale gas boom) and into Asia after the March 2011 Fukushima disaster.

Spot trading accounted for 25% of LNG sales in 2011. This number should continue to grow as recent contracts allow rerouting to take place.

– Tankers freight rates

Any sudden change in supply (disruption) or in demand (like the nuclear phase out) creates relaxation or tightness in the shipping industry, making tankers freight rates highly volatile. Price of spot cargo has historically been between 30,000 $/day and 160,000 $/day. Some ships have been built for speculative purposes as owners expect that any sudden change in the LNG balance could tighten the shipping industry and send freight rates back to their record high level.

Figure 78

Costs along the LNG chain.
Source: SG Cross Asset Research.

– Arbitrage costs more than 1 $MBtu

When carried on ships, LNG tends to revaporise (or boil off). The boil-off is used to propel the ship: with modern technology it can make up to 90% of the energy needed for the propulsion. The cost of LNG used to price the boil-off is between 2 $/MBtu (cost of LNG in the case of a national oil producer) and 4 $/MBtu (price of LNG sold at the exit of the liquefaction plant). In our example, we use an average of 3 $/MBtu.

Table 4 Arbitrage costs

Size of the tanker		a	150,000	cm of LNG
		b=600*a	90,000,000	cm of gas
		c=b*35.31/1000	3,177,900	MBtu
Speed		d	19	knots
Boil-off		e	0.15	% per day
Long term freight rate		f	80,000	$/day
Estimated cost of LNG (used to price the boil off)		g	3	$/MBtu
Egypt - Japan (1 way)		h	8,000	miles
		i=h/(24*d)	18	days
Cost of Freight rate		j=2*f*i	2, 807,018	$
Cost of boil-off		k=2*e*i*g	501,774	$
LNG left for sale		l=c-j	2, 676,126	MBtu
Approximate cost of shipping from Egypt to Japan		m=(j+k)/l	1.24	$/MBtu
Egypt - UK (1 way)			3,000	miles
			7	days
Cost of freight rate			1,052,632	$
Cost of boil-off			188, 165	$
LNG left for sale			2,989,735	MBtu
Approximate cost of shipping from Egypt to UK			0.42	$/MBtu
Approximate extra cost to divert Egyptian LNG away from UK to Japan		(excl. Suez Canal)	0.82	$/MBtu

The shipping time doesn't take into account the loading and unloading of the ship. Each operation takes 36 hours on average, except for ships with onboard regas where it can take up to a week.

On top of this, the price of going through the Suez Canal has to be added when the ship takes this route. A return trip costs c.$1 m or 0.37 $/MBtu. This shows that canal tolling is expensive and calculated in function of the use of the canal. This is why when the Panama Canal will be enlarged, in 2014e, the tolling for LNG carriers will be calculated in function of the number of days this canal is allowing to spare.

The basic calculation above shows that diverting a 150,000 cm tanker will cost 0.8 $/MBtu + 0.4 $/MBtu for the Suez tolling; the total cost of arbitrage is therefore 1.2 $/MBtu. On the same basis, Trinidad & Tobago LNG redirected away from the US to Japan will take 29 days and cost 1.1 $/MBtu (with no tolling).

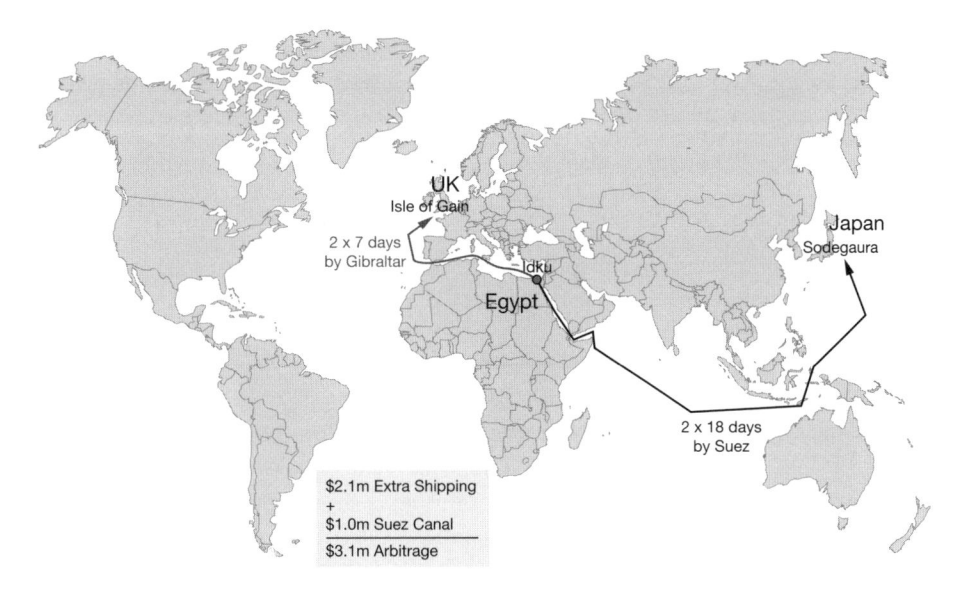

Figure 79

East or West? A $3.1m question.

– How much can an arbitrage earn?

As seen above the extra cost of a tanker arbitrage is c.$3.1m. Traders in this market are prepared to take an arbitrage risk if the rewards are substantial, i.e. the price differential has to be above 1.7 $/MBtu (1.2 $/MBtu for the cost of arbitrage and 0.5 $/MBtu for the minimum reward or $1.3m for the full ship). Arbitrages can earn on average $10m (with a price differential of 4.9 $/MBtu in our example), with very good arbitrages reaching $30m (with a price differential of 12.4 $/MBtu in our example).

– Regas capacity: a financial option

Faced with a low regas capacity load factor (35% on average), market power is likely to shift further into the hands of producers, unless we view regas terminal as a financial option to allow arbitrages.

Although regas rental charges are confidential, if a terminal was fully used, the charges taken should equal 0.5 $/MBtu. To deliver 1.3 bcm (1 mt) of LNG, 11 of our standard ships must berth. To cover the cost of renting 1.3 bcm/y (1 mtpa), a company must be able to make $24m. This can either be viewed as a necessary cost for a company with secure supply (producers or companies with long-term contracts) or a speculative margin for a company wanting to play LNG arbitrages. For the latter option to be successful the investor needs either one successful arbitrage or 11 average arbitrages with a $2.2m gain each. It is therefore obvious that some terminals are operating at such a low capacity that users don't even cover their cost of renting! However, given that they had the molecules they took the risk of renting regas on a long-term basis to be able to extract more value from their resources by rerouting their cargoes to higher markets, if needed.

LNG producers and aggregators will be able to maximise their profits by always benefiting from spikes. A severe disruption (nuclear outages, droughts, hurricanes, etc.) would likely send consumers in a mad scramble for the last available LNG carrier. LNG prices would spike as a result and the companies with the supply and trading expertise just have to hold on to enjoy the huge benefits that should pour in from time to time.

– Ripple effect: LNG prices could even drive piped gas

By constructing too many regas terminals (most of them not backed by long-term contracts for LNG supply) in consuming countries we see risks of a global fight for the marginal cargo. Every country with a regas terminal is showing readiness to embark on a global race for the most expensive marginal molecule. In a global market, a customer with LNG and pipe import facilities could, in theory, arbitrage between the cheaper. In a tight market, the customer tends to take what is available even at the highest price. The piped gas supplier could be tempted, in the future, to align its marginal price not to the regional price but to the global, higher price. But in a relax market this could have the opposite effect.

While the LNG market is indeed growing, it represents only 9% of consumption and would need to expand significantly through massive amounts of investment in order to fully develop a functioning spot market. For LNG to fill this greater role, a number of obstacles must be overcome. The market limitation is simply in the amount of production facilities and we are unlikely to see a significant increase in these facilities in the next few years. In the future, however, LNG may very well be the price setter for gas. This could have an impact on piped gas as well.

MARKETS CAN SELECT THE MOST PROFITABLE FUEL TO GENERATE ELECTRICITY

As we've seen earlier, the merit order to generate electricity is first solar and wind when available, baseload nuclear, cheap hydro and then gas, oil or coal depending of their respective pricing.

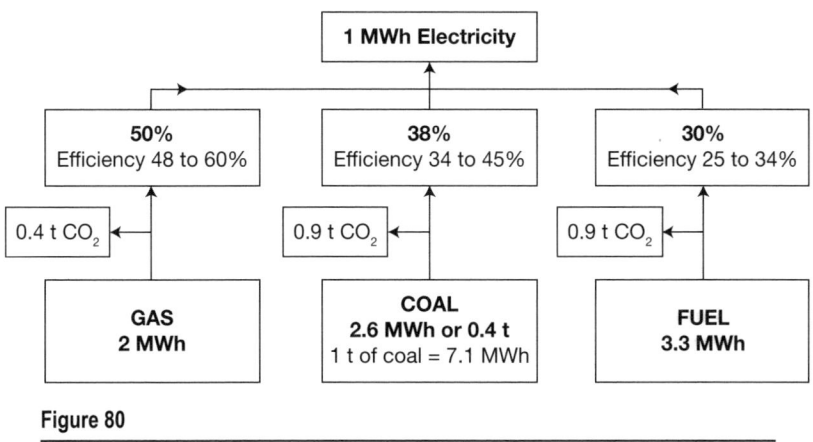

```
                        1 MWh Electricity
         ┌──────────────────┼──────────────────┐
      50%                  38%                30%
  Efficiency 48 to 60%  Efficiency 34 to 45%  Efficiency 25 to 34%
       ↑                    ↑                   ↑
   0.4 t CO₂            0.9 t CO₂           0.9 t CO₂
       ↑                    ↑                   ↑
      GAS                  COAL                FUEL
     2 MWh           2.6 MWh or 0.4 t        3.3 MWh
                    1 t of coal = 7.1 MWh
```

Figure 80

Overview of electricity generation from fossil fuels.

The burning of coal, gas or oil to generate power, also generates CO_2, a pollutant that needs to be reduced if we want to combat climate change. The inter-fuel competition needs to take into account not only the cost of gas, coal or fuel oil but also the CO_2 price.

In Europe, gas and coal are in competition to generate electricity. Gas is a cleaner fuel but this doesn't mean gas is going to be selected instead of coal. The market is selecting gas or coal depending on the profitability of each plant.

In technical terms, traders make the distinction:

• Spark spread (price of 1 MWh of electricity – price of 2 MWh of gas) represents the margin of an average gas fired power plant without taking into account the CO_2 emissions.

• Dark spread (price of 1 MWh of electricity – price of 0.4 t of coal) represents the margin of an average coal fired power plant without taking into account the CO_2 emissions.

• Clean spark spread (price of 1 MWh of electricity – price of 2 MWh of gas – price of 0.4 t of CO_2) represents the margin of an average gas fired power plant taking into account the CO_2 emissions.

- Clean dark spread (price of 1 MWh of electricity – price of 0.4 t of coal – price of 0,9 t of CO_2) represents the margin of an average coal fired power plant taking into account the CO_2 emissions.

Traders adapt production to maximise gain, hence if clean dark spread is higher than clean spark spread, the coal-fired plant will be selected and the gas-fired one will stay idle.

Gas is a greener fuel but markets are not selecting fuels for power generation according to their environmental merit but to their profitability. Trends across Europe of low profitability for gas-fired power plants are signalled by low spark spreads. If we want markets to select gas, then the pricing of CO_2 must push for this.

For example, in the UK, since April 2011, it has been more profitable to burn coal than gas (priced on NBP). The market says that utilities should use first coal fired plant and if this is not enough then they should turn to gas.

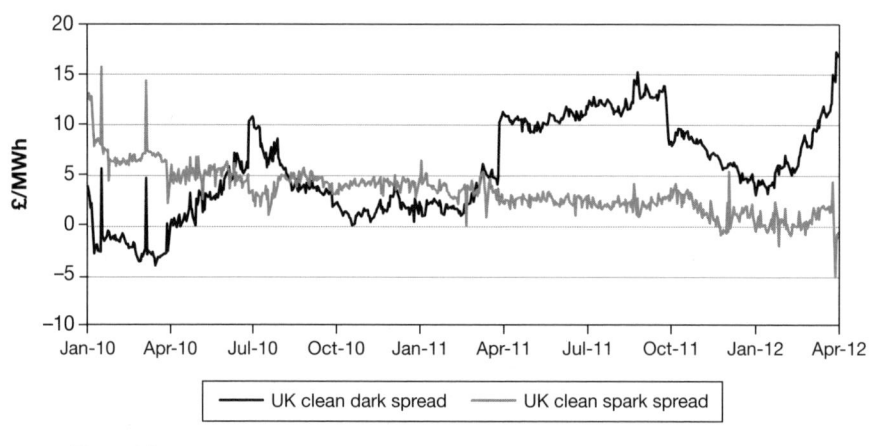

Figure 81

UK Next Season Clean Spreads.
Source: SG Cross Asset Research, Reuters.

We will look at this more in depth when looking at European policies as we need first to explain how CO_2 is priced in Europe.

MARKETS INTEGRATION?

Today we have 3 separate gas markets: US (spot), Europe (oil-indexation and spot) and Asia (oil-indexation). In theory, non-contracted LNG should flow to the highest priced market and we should see some integration. This is not yet the case as:

- The US is a "gas island" that doesn't need any LNG and therefore is completely disconnected from the rest of the word.

- LNG accounts for only 31% of total gas exported (except intra-CIS trade) and most of it (c.75%) is sold under long term oil-index contracts, with little rerouting allowed.

As we will see later, as Europe moves to mainly spot indexation and if we see the US becoming a major LNG exports (with non oil-index LNG contracts), world-wide gas markets could integrate. Markets would then be linked via the cost of arbitrage.

Policies

— The difficult task of balancing cheap, secure and clean energy

Until the shale gas revolution, policy was concentrated on prices, security of supply and clean energy. The question of dependency was not put forward as net importers were assuming that their oil & gas dependency was only supposed to increase. We will see later that the shale gas revolution is in fact offering a lot of net importing countries the opportunity to completely review this dependency issue.

Table 5 Comparison between different fuels to generate electricity

	Cheap	Secure	Clean
Oil	x	x	x
Gas	x	x	✓
Coal	✓	✓	x
Renewables	x	✓	✓
Nuclear	?	?	?
Domestic shale gas	✓	✓	✓

If we don't take into account the production side of fossil fuels and nuclear extraction, domestic shale gas ticks all the boxes: it is the only fuel that is cheap, secure and clean…

CHEAP ENERGY ALLOWS DEVELOPMENT

Steam engines ignited the Industrial Revolution (1750 to 1850) in the UK and propelled it around the world…

The industrial revolution started with the mechanisation of textile industries, the development of iron-making techniques and the increased use of refined coal. The introduction of steam power fuelled primarily by coal, wider utilisation of water wheels and powered machinery underpinned the dramatic increases in production capacity. It began in the UK (Manchester) and subsequently spread throughout Western Europe, North America, Japan, and eventually the rest of the world.

... And the opposite is also true; with high oil prices, recession is always looming as we've seen in the 70's, in 2009 and 2012.

SECURE ENERGY HAS BEEN THE FIRST POLICY GOAL AS IT IS ESSENTIAL TO NATIONAL SECURITY

Fortunately, the market does a remarkably good job in achieving day-to-day security. In oil, where the paper barrel is traded 50x more than the physical barrel is consumed, market can reallocate barrels to make sure we continue to have petrol for our cars. In gas, hubs are also making sure lights stay on.

As consuming states are afraid of energy supply risks, the most obvious policy goal is "Security of Supply" (laws, European directives, international agreements, etc. are designed to make sure citizens should always get access to energy). But it is more difficult to mitigate a supply risk of gas (particularly in winter) than of oil. It is also fair to mention that when consuming states are worried about Security of Supply, producing states are looking for Security of Demand as they want their investment in E&P to be profitable. Behind the word "secure" also lies the "dependency" issue. Until the shale gas revolution net importers were bond to become more and more energy dependant (and net exporter could also face the "Dutch curse", the economic phenomenon in which the revenues from natural resource exports damage the nation's productivity). As we will see later, the shale gas revolution changed this dependency paradigm forever and is offering an alternative...

• In Oil, Strategic Stocks are a Reality and have been used

Founded during the oil crisis of 1973-74, the International Energy Agency (IEA) initial role was to coordinate measures in times of oil supply emergencies. Today IEA acts as energy policy advisor to its 28 member countries in their effort to ensure reliable, affordable and clean energy for their citizens.

The basis for the emergency policy lies within the treaty upon which the IEA was founded in 1974. The treaty includes the commitment of IEA member countries to hold oil stocks equivalent to no fewer than 90 days of the prior year's net imports. The treaty also includes an integrated set of emergency response measures for major international oil disruptions, with the following emergency response measures:

- Drawdown of oil stocks.
- Demand restraint.
- Fuel-switching out of oil.
- Surge oil production.
- Sharing of available supplies.

Those measures were effectively used during the first Gulf War (1991) when the IEA implemented a 2.5 mb/d contingency plan, most of which was stock draw. In September 2005, the IEA implemented a collective action, which made available to the market 60 mb of crude oil and oil products, in response to concerns about interruptions to oil supply as a result of the severe hurricane damage caused by Hurricanes Katrina and Rita in the US Gulf of Mexico. Most recently, in June 2011, IEA agreed to release 60 mb of oil in response to the disruption from Libya.

– Another system, at EU level, to make things difficult to understand!

Prior to the IEA treaty, in 1968, a directive imposed an obligation on member states of the then EEC to maintain minimum stocks of crude oil and/or petroleum products. In 1998, the directive was amended and today EU member states must have minimum stocks of petroleum products at a level corresponding, for each of the categories of petroleum products, to at least 90 days' average daily internal consumption in the preceding calendar year (internal consumption met by indigenous petroleum production may be deducted). And on top of the accounting difference, the EU has a requirement to stock products when the IEA doesn't.

• In Gas, Strategic Stocks are introduced

Unlike the case of legislation on strategic oil stocks, IEA and EU don't impose any obligation on member states to maintain gas strategic stocks.

The European Commission strongly encourages all member states to support the development of commercial storage, but has not proposed an EU-level obligation regarding strategic stocks. However, EU member states are allowed to impose public service obligations which may relate to supply security.

Mandatory gas stocks and/or strategic stocks were introduced in Italy, Hungary, Portugal, Spain and Poland. It is understandable that the non-producing countries with the higher gas share in their primary energy mix (Hungary and Italy) were the first to decide to go for gas strategic storage as they are the most vulnerable to a supply crisis.

The January 2009 gas crisis revealed serious shortcomings in the security of supply in South Eastern Europe. There were no gas supplies from Russia to Europe from the 7th to 20th of January 2009. The gas supply disruption lasted a total of 14 days. It took almost as long for emergency network options, such as reverse flows, to be implemented where interconnections where already in place (Czech Republic to Slovakia and Greece to Bulgaria). Reverse flow proved a problematic option at the time and came too late to make a real difference, particularly in Bulgaria. Reverse flow was recognised as potentially powerful tool for mitigating the effects of supply disruptions. However, more investment in the relevant infrastructures was needed to ensure that this possibility exists across the EU's internal market. The crisis also served as a wake-up call to policy makers to strengthen their energy strategies. The crisis created momentum for better regional cooperation and dedicated capex was spent to allow gas to flow reverse from West to East if needed.

There is a clear need for more investments in storage, LNG, technical equipment to reverse flows and interconnections between Member States and countries outside the EU in order to increase flexibility and mitigate potential supply disruptions. The realisation of these projects which can substantially enhance the flexibility and security of gas supply and better interconnect all EU Member States has already started. In 2010/11 the European Energy Programme for Recovery (EEPR) supported the construction of 31 gas infrastructure projects with €1.4bn. Learning from the lessons of the January 2009 gas crisis, the EEPR supported projects for reverse flow in 9 Member States with around €80m and gas interconnectors with around €1.3bn, including new import pipelines.

– 5.1 bcm in Italy

In 2001, a law set a requirement on shippers in Italy to have strategic gas stocks of 10% of non-EU imports. This translated then into 5.1 bcm of strategic stocks. This "national insurance policy" is imposed on shippers importing gas produced from non-EU countries.

– 0.9 bcm in Hungary

Gas is the most important fuel in Hungary, representing 42% of the country's total primary energy mix (vs. 26% in Europe). The share of gas in the household and public sector heating exceeds 70%. Following the supply interruption of January 2006, the Hungarian parliament approved a law on safety stockpiling of natural gas in February 2006. A new underground storage with a total capacity of 1.2 bcm was constructed and operational by 2010. In June 2010, Hungary amended the legislation to allow for a reduction in the minimum strategic stockholding level, the level of which is to be determined on a yearly basis by the Minister. As of May 2011, the country's strategic stock level is 0.9 bcm.

– *Minimum stocks in Poland, Portugal, Spain and Ukraine*

Every undertaking which is shipping natural gas to Poland or supplying customers on Polish territory with gas not originating from domestic production is obliged to keep strategic storage of gas. The amount of natural gas to be stored equals the volume of natural gas shipped to Poland during an average 30 days of its operation. The level of 30-days reserve of natural gas will be required starting October 1st 2012.

In Portugal, gas importers are mandated to hold gas reserves of 15 days' consumption of non-interruptible gas-fired power plants and 20 days' consumption of non-interruptible customers of the remaining market. The estimated 0.2 bcm gas inventories that may be accounted for the purpose of mandatory security reserves are the combined existing stocks of each responsible market agent at the underground storage, at the LNG storage, and at LNG carriers with fixed port destination in Portugal, with an estimated time of arrival of up to 9 days.

In Spain, operators are obliged to maintain minimum stocks equivalent to 20 days of consumption ("strategic reserves"). The gas sector obligation is calculated in accordance with what the regulations define as "firm sales" (supplies that cannot be interrupted, either for commercial or technical reasons) during the preceding calendar year. The stocks corresponding to this obligation must be maintained between April 1st and March 31st of the following year.

There is also a regulatory obligation in Ukraine for suppliers to store 10% of their annual supplies.

CLEAN ENERGY, IF WE HAVE THE MONEY!

The more affluent a society becomes, the more willing it is to pay for clean energy. Since these changes were the result of consumer choices, at first no regulation was needed. One of the best examples is the displacement of coal in favor of gas to heat houses in big cities during the first part of the XX century. London's famous fog that can still be admired is some of Turner's (1775-1851) and Monet's (1840-1926) finest paintings (respectively *The Scarlet Sunset* and *Charing Cross Bridge*) has receded thanks to the fact that coal is not used as a heating fuel any longer.

For many developing countries, clean energy is a luxury they cannot yet afford; their priority remains cheap energy. China is in the process of securing its energy (by building strategic oil stocks, by investing in hydrocarbon production abroad, by signing long term gas contracts, etc). Big Chinese cities (where the wealthiest are leaving) are also looking at cleaner air by forcing coal-fired plants (power or industry) in their vicinity to close down or to switch fuel. Presumably as income rises,

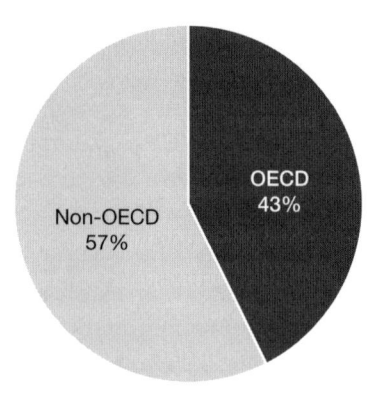

Figure 82

2010 split of CO_2 emissions between OECD and non-OECD.
Source: BP Statistical Review.

developing countries will be increasingly concerned with air pollution and will demand cleaner fuels. Unfortunately, it may take many more years. In the interim, rapidly rising greenhouse-gas emissions pose a worldwide problem...

China, the world's biggest user and producer of coal, will limit consumption of the commodity to reduce pollution and curb reliance on the fuel, according to a five-year plan for the coal industry released by the National Energy Administration, in March 2012. China, the world's biggest producer of carbon emissions, will strengthen control of air pollution. The government aims to cut carbon dioxide emissions by as much as 17% per unit of gross domestic product in its five-year plan through 2015.

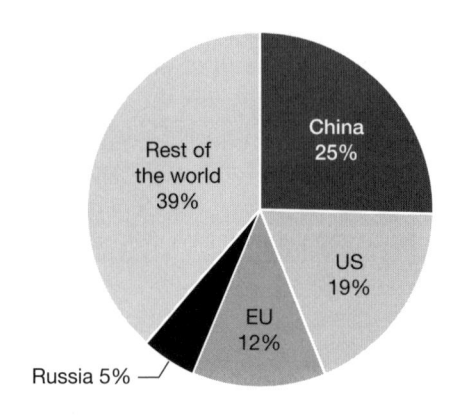

Figure 83

2010 top CO_2 polluters.
Source: BP Statistical Review.

• EU ETS: too Complex

The EU Emissions Trading System (EU ETS) is a cornerstone of the European Union's policy to combat climate change and its key tool for reducing industrial greenhouse gas emissions cost-effectively. Being the first and biggest international scheme for the trading of greenhouse gas emission allowances, the EU ETS covers 30 countries (the 27 EU member states plus Iceland, Liechtenstein and Norway). It covers CO_2 emissions from installations such as power stations, combustion plants, oil refineries and iron and steel works, as well as factories making cement, glass, lime, bricks, ceramics, pulp, paper and board.

Launched in 2005, the EU ETS works on the "cap and trade" principle. This means there is a "cap", or limit, on the total amount of certain greenhouse gases that can be emitted by the factories, power plants and other installations in the system. Within this cap, companies receive emission allowances which they can sell to or buy from one another as needed. The limit on the total number of allowances available ensures that they have a value.

At the end of each year each company must surrender enough allowances to cover all its emissions, otherwise heavy fines are imposed. If a company reduces its emissions, it can keep the spare allowances to cover its future needs or sell them to another company that is short of allowances. The flexibility that trading brings ensures that emissions are cut where it costs least to do so.

Before the start of the first (2005-2007) and the second (2008-2012) trading periods, each member state had, in line with the framework of the Kyoto Protocol, to decide how many allowances to allocate in total for a trading period and how many each installation covered by the ETS would receive. For the third trading period, which begins in 2013, there will no longer be any national allocation plans. Instead, the allocation will be determined directly at EU level.

Airlines should join the scheme in 2012, but the outcome is uncertain as the dispute is intensifying over this new "tax". The EU ETS will be further expanded to the petrochemicals, ammonia and aluminum industries and to additional gases in 2013, when the third trading period will start. At the same time a series of important changes to the way the EU ETS works will take effect in order to strengthen the system.

In the EU ETS, the number of allowances is reduced over time so that total emissions fall. In 2020 emissions should be 21% lower than in 2005. But it is losing credibility as:

• The recession made the targets achievable.

• UK plans to reward low-carbon electricity generation, which includes nuclear power. UK power producers burning fossil fuels like gas and coal will have to pay the government surcharge, which is calculated two years in advance, in addition to buying EU permits to cover emissions. The so-called carbon price support is set at

4.94 £/t for 2013, when the policy takes effect, moving to 9.55 £/t for 2014. The price support is intended to bridge the gap between EU permit prices and the UK's carbon floor, which has been set through 2020 and determines the minimum level industries will pay to emit. The carbon floor starts at 16 £/t in 2013, rising to 30 £/t by 2020.

• The 1997 Kyoto Protocol, which called for its signatories to meet certain greenhouse-gas-emissions target in 2012, is losing steam with Canada decision to exit it in December 2011. What happens beyond 2012 is one of the key issues governments are currently negotiating.

• Carbon Capture and Sequestration (CCS) is actually not feasible on an industrial level and according to experts this should cost North of 80 €/t, far above today trading level of CO_2 (7 €/t). This shows that the EU ETS is not providing a price signal that can push for CCS technologies to be implemented.

Climate change is a complex problem, which, although environmental in nature, has consequences for all spheres of existence on our planet. Consumers exhibit little willingness (or ability) to reduce their use of energy in the short run in response to price increase as they are stuck with their existing equipment infrastructure. In the long run, however, the response is quite different. As over time, consumers can replace their energy-consuming equipment with more efficient cars and other appliances.

The actual ETS is too complex and citizens have been left out. On top too many lobbies and too much regulation are making the system more and more complex. We believe either the trading principle is kept and simplified or a simple tax that is easily explainable (even if nobody likes taxes!) could be introduced. The case for moderate initial tax that rises substantially over time is compelling. The existing stock of plants, vehicles and buildings was configured on the basis of past energy prices and, in the short term, there is little one can do other than drive and heat less. Knowing that the carbon tax will increase dramatically over a long term period, consumers would begin altering their investment decision today.

A carbon tax:

• Establishes an observable price society is willing to pay for CO_2 abatement,

• Avoids the potential problem of uncertain and highly fluctuating prices of emissions rights. Investors in new plants know the current carbon tax and the rate at which it will escalate in the future and thus can formulate sensible long term investment plans.

• Allows filling states coffers at a time where high debt burden is a reality.

• Would (with no exemptions) be much more transparent than tradable emissions allowances and less subject to manipulations.

• Would be much simpler than the ETS.

As we don't think that CCS should be available before 2025e, the tax could be assessed on the carbon content of the fuel, which is readily measurable. Credits for carbon sequestration technologies would be allowed only on the basis of measured quantities of CO_2 effectively sequestered, when the technology gets available.

The carbon tax will raise the price of fossil fuels to reflect the true costs of such fuels. The higher prices will make alternative fuels more economically attractive and induce more energy conservation.

Setting the price of CO_2 (either by a trading mechanism or a tax) is very important in the case of power generation. The EU ETS was designed to allow market to find the "greener route"… But traders adapt production to maximize gain and look at the clean dark and clean spark spreads to select which plants to put on-line first.

In Spain, a new domestic coal support law went into effect in February 2011. Coal-fired plants are displacing combined-cycle gas plants as domestic coal generation is forced into the system. In 2011, gas demand for electricity was reduced by 19%…

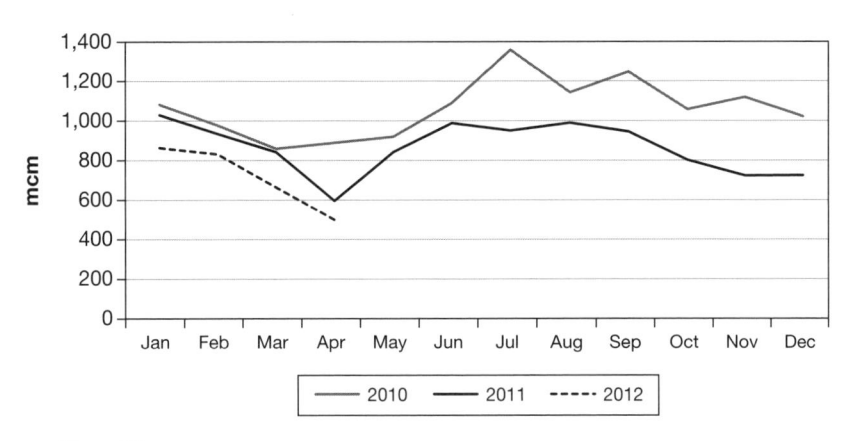

Figure 84

Gas demand for electric generation in Spain.
Source: SG Cross Asset Research, Enagas.

TAXES / SUBSIDIES

As energy is strategic for states, specific taxation applies both on the producing and consuming side. We've already seen the small mineral extraction tax in Russia (that has nevertheless been increased a lot in recent years) and the export duty tax. Those taxes can be adapted. For example, the Russian export duty tax doesn't apply to:

- Russian gas flowing in Turkey via the Blue Stream thanks to a Turkey-Russia intergovernmental agreement.
- Gazprom's exports of non-Russian gas (i.e. Caspian gas that flows via the Russia grid to Europe).
- Exports of Russian gas to Customs Union countries (Belarus).

Difficult projects also need very favourable tax regimes (tax breaks, low mineral extraction tax, reduced export duty, etc.) to be profitable. This has to be decided by the producing state for a long period of time allowing investors to make long term plans without the risk of possible changes if the state believes market conditions make it expedient.

In Norway, the corporate tax is 28%. In addition, the petroleum tax act imposes a special tax of 50% on income from offshore oil exploration, development, processing, production and pipeline transportation. In total, Norway levies a 78% tax rate.

In 2011, the UK government surprised the industry by increasing the effective tax rate on offshore output from 50% to 62%, made up of 30% corporation tax plus a 32% supplementary charge.

States are also tempted to subsidise energy to please their citizens. This applies all over the world. In Russia, until very recently, regulated prices for residential customers were not covering the full cost of production. In France, before each election, the government is trying to freeze regulated gas prices for residential customers. In Poland, PGNiG has been incurring losses due to delayed regulated price increases...

Addressing fuel poverty (when someone has to spent more than 10% of his income on fuel for heating and cooking) is a major challenge even in OECD countries... For example, Britain has a legally binding target to eradicate fuel poverty as far as reasonably practicable by 2016, but all studies are predicting that this target is over ambitious.

Subsidies are cancelling the price signal. When energy prices are regulated below international prices (or below full cost of production) then clients are not incentivised to reduce their consumption and gas companies profits are negatively impacted by this, reducing their ability to invest. Too low regulated prices lead to wasteful consumption.

Given current technologies, renewable energy (solar and wind) is substantially more expensive for power generation than fossil fuels. Therefore, power generation from renewables would not be viable without government support through subsidies. Those subsidies distort the fuel mix in favour of green energies. And once the capex is spent and the solar panels and/or wind farms are connected they continue to generate electricity, even if the subsidies are reduced.

REGULATION / STATE INTERVENTION

As energy is strategic, states need to regulate this industry or want to intervene. Just a brief overview of what can be found...

• In 1938, the US government first regulated the natural gas industry. At the time, members of the government believed the natural gas industry to be a "natural monopoly". Because of the fear of possible abuses, such as charging unreasonably high prices, and given the rising importance of natural gas to all consumers, the Natural Gas Act was passed. This Act imposed regulations and restrictions on the price of natural gas to protect consumers. Into the 1980s and early 1990s, the industry gradually moved toward less regulation; the beginning of the XXI century has brought significant regulation. Today, the natural gas industry is regulated by the Federal Energy Regulatory Commission (FERC).

• Just after the end of World War 2, in 1946, in France, gas and electricity were viewed as essential and the two companies were therefore nationalised.

• Policy interventions can range from moratorium (as we've seen on Qatar) to production cap in the Netherlands. The Groningen field concession belongs to NAM (50% Shell and 50% ExxonMobil). "Maatschap Groningen" (60% NAM and 40% Dutch State) is in charge of producing this field. The Dutch law authorizes for the 2006-2015 period, a maximum production of 425 bcm (an annual average production of 42.5 bcm) to preserve this field for future generations.

• Turkey plans to reduce its gas power generation from 50% today to 30% in 20 years, to be less dependent on gas importers.

But the tendency to over regulate this industry by adding, with time, new layers of laws and/or regulations, can also be a risk. Most failures, be it market or policy related, lead to new regulations...

• Focus on European Storage Regulation

As incumbents use the best available geological structures to create their storage, historical storage operators should, in a liberalised gas market, get a huge premium. To prevent this, with the opening of the gas market, different regulations were set in Europe. The 2003/55/CE European directive (of 26 June 2003) imposed third-party access (TPA) to storage (exceptions can be granted on the basis of article 22). Non-TPA also includes strategic stocks (in Italy and Hungary).

Member states could choose between a regulated (tariff set by an independent regulator), or a negotiated (tariff set by storage operators) access. In a negotiated environment, the storage operator can try to extract market value from its asset, while in a regulated environment, the economic margin sets by the regulator is between 6 and 16%.

Table 6 Different TPA regimes in Continental Europe

Regulated	Negotiated
Belgium	Austria
Hungary	Czech Republic
Italy	Denmark
Latvia	France
Poland	Germany
Portugal	Netherlands
Romania	
Slovakia	
Spain	

Source: SG Cross Asset Research, GIE.

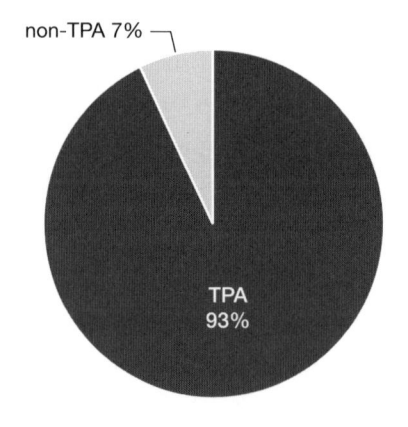

Figure 85

Split of European storage: TPA and non-TPA.
Source: SG Cross Asset Research, GIE.

This explains why governments and regulators have a say when it comes to storage.

- In April 2012, the Spanish government decided to halt the start-up of the 1.3 bcm Castor offshore gas storage site due weak gas demand to avoid increasing the costs of operating the national gas system. The site was scheduled to start in 2013. It will be postponed until there is a balance between supply and demand on the market.

- Longer-term storage will continue to be an important method of ensuring security of supply and the UK government is still pushing for more storage capacity (as it could be too dependent on LNG swings as discussed later). But, in the UK, the market has so far decided against it!

Where is the Future Supply Growth?

THE US SHALE GAS REVOLUTION

– Shale gas was a driving force in helping the US to overtake Russia as the world's largest gas producer in 2009. Unconventional gas now allows the US to enjoy much lower gas prices than any other country. The US shale gas revolution has transformed the US into a gas island.

According to the US DoE, total marketed gas production grew by 7.9% in 2011, the largest year-over-year volumetric increase in history. This strong growth was driven in large part by increases in shale gas production. In 2012, declines in production have not accompanied declines in the rig count, partly reflecting improving drilling efficiency. That fact, combined with high initial production rates from new wells, associated natural gas production from oil drilling, and a backlog of uncompleted or unconnected wells contribute to the US DoE's forecast of further production increases in 2012 and 2013. And in Q1 12, production grew by 8.8% vs. Q1 11.

According to the US DoE, total gas consumption grew by 2.5% in 2011. US DoE expects that natural gas consumption will further increase in 2012 and 2013. Consumption increases in all sectors, with the largest volume increase (c.60% of total growth) coming from the electric power sector as increases in the consumption of natural gas for power generation are likely to continue as domestic production continues to grow and natural gas remains a relatively inexpensive option for generators.

But without any new uses (transport?) or export facilities, the US "gas island" is facing oversupply. This explains recent decisions to try to curtail gas production.

Chesapeake decided, in January 2012, to curtail c. 5 bcm/y, or 8%, of its current operated gross gas production of 63 bcm/y (which is about 9% of the US natural gas production). If conditions warrant, the company is prepared to double this production curtailment to as much as 10 bcm/y and plans to defer new dry gas well completions and pipeline connections wherever possible in response to low US natural gas prices. Chesapeake's idea is to redirect capital savings from curtailing dry gas production to its liquids-rich plays that deliver superior returns.

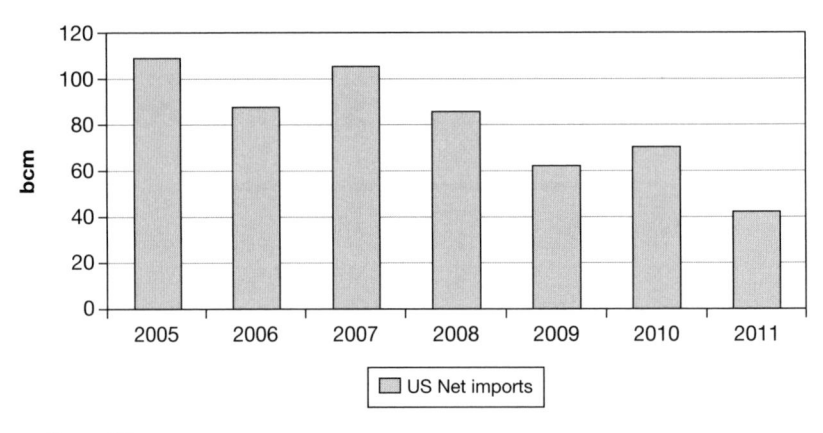

Figure 86

US net imports.
Source: BP Statistical Review, US DoE.

This decision to redirect capex away from US dry shale gas production was then followed by other major companies (ConocoPhillips, Statoil, Shell and BG) as actual US prices are too low for companies to be profitable.

Whatever the future of US gas prices (from 2 to 7 $/MBtu) it would be cheap unless fracking is banned for environmental reasons. Fracking gas from tight rock formations promises energy supplies for generations, but only if industry and regulators can convince voters it can be done safely without poisoning water supplies or adding to global warming... All over the US, regulators are reviewing and strengthening all rules related to shale production (disclosure of chemicals, water recycling, air quality, etc.) to include unconventional oil & gas in the regulatory framework of each producing state.

Shale gas production in the US increased by a CAGR of 46% over 2005-2010 to account for 23% of 2010 US dry gas production.

The US DoE expects shale gas to account for 49% of US dry gas production by 2035e.

– As the number 1, worldwide gas producer can the US change the geopolitics and reduce Russia's power on the energy scene?

Shale gas production has transformed the US gas industry in recent years, boosted production rates and booking reserves of gas as we've seen in the chapter "Basics". Thanks to the shale gas revolution, the US has become a low cost energy producer on a global basis. However, opponents of shale gas extraction in the US have criticized the industry for not disclosing the chemicals included in fracking fluids, which they say could contaminate water supplies. Some companies claimed

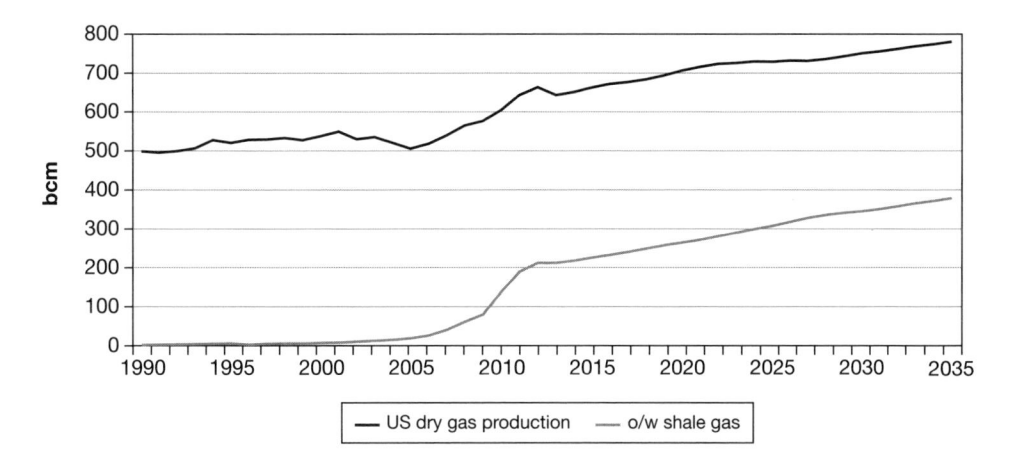

Figure 87

US yearly gas production.
Source: US DoE.

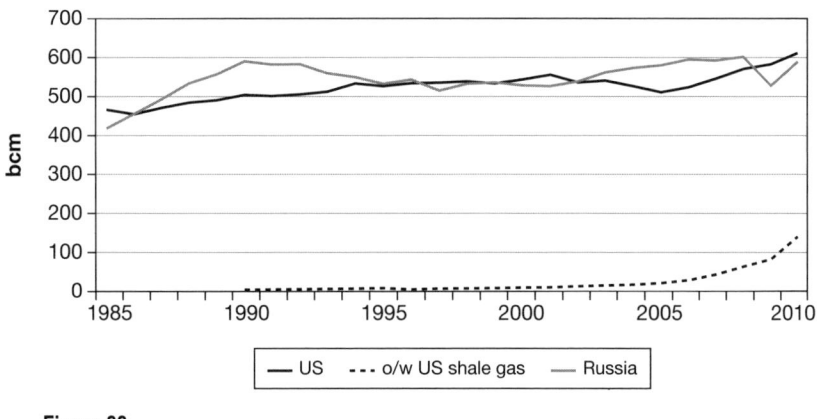

Figure 88

US shale gas changed worldwide ranking.
Source: SG Cross Asset Research.

that the details are commercially sensitive. The fracfocus website (fracfocus.org), an online registry of fracturing chemicals, allows citizens to access the information concerning the materials used to fracture the well. Only transparency can improve acceptability of this technology.

– Continued growth in production should enable the US to become an LNG exporter from as early as 2016e.

Recent developments in the US and Canada could lead to North America becoming a major LNG exporter. For a liquefaction facility to be built in the US, a wide range of authorisations are needed:

• An important one is granted by the Department of Energy (DoE) to allow exports as any state has permanent sovereignty over its natural resources (UN Resolution 1803 (XVII) of 14 December 1962). An application for export authorisation has to be filed by companies that want to build and operate an LNG export terminal. The DoE can grant authorisation either to countries with which the US has a free trade agreement (FTA countries are Australia, Bahrain, Canada, Chile, Costa Rica, Dominican Republic, El Salvador, Guatemala, Honduras, Israel, Jordan, Mexico, Morocco, Nicaragua, Oman, Peru, Singapore and South Korea. Colombia and Panama should join the FTA countries once all legislation is passed) or to all countries with which trade is not prohibited by US law.

• Another one is granted by the FERC to site, construct and operate facilities for the liquefaction and export of domestically produced natural gas. This process takes more than a year and costs a few million $ as a lot of studies are needed.

In less than 4 months, Cheniere, that was the first company to be granted a DoE authorisation to export US LNG to FTA and non-FTA countries, managed to sell all its LNG (16 mtpa) under a Henry Hub linked formula (LNG delivered Free On Board: 115% HH + fixed fee). The 115% HH cover the gas sourcing (100% at the hub), the cost of fuel gas needed for the process (10%) and additional transportation cost to the liquefaction terminal (5%). The fixed fee is for the remuneration of the liquefaction plant that will therefore operate as a tolling plant. Buyers have the right not to take the LNG as long as they pay for the fixed fee. BG will purchase under a 20-year LNG Sale and Purchase Agreement 3.5 mtpa (4.7 bcm/y) of LNG from Sabine Pass (Louisiana) with the commencement of train 1 operations and will purchase a portion of the additional 2 mtpa (2.7 bcm/y) of LNG as each of trains 2, 3 and 4 commence operations. Gas Natural Fenosa, Kogas and Gail will purchase each 3.5 mtpa (4.7 bcm/y) from respectively trains 2, 3 and 4. In October 2011, BG managed to get the lowest tolling price (2.25 $/MBtu) to reflect BG's status as a foundation customer. Gas Natural Fenosa, in November 2011, did get 2.49 $/MBtu and GAIL and KOGAS, respectively in December 2011 and January 2012, 3 $/MBtu. In January 2012, for the additional LNG, BG will pay a tolling price of 3 $/MBtu, showing that Cheniere has more market power as its Sabine Pass project is now viewed as the one that could be the first to deliver US LNG. With 86% of phase 1 (trains 1 & 2) contracted, Cheniere has secured an annual revenue of at least $975m for an investment that should be $5bn (before financing costs). As Cheniere has been granted all authorisations, US LNG should arrive as soon as 2016e thanks to Sabine Pass.

In April 2012, Cameron LNG signed commercial development agreements with Mitsubishi and Mitsui to develop and construct a liquefaction export facility in Louisiana. The commercial development agreements bind the parties to fund all development expenses, including design, permitting and engineering, as well as to negotiate 20-year tolling agreements, based on agreed-upon terms outlined in the commercial development agreements. Each tolling agreement would be for 4 mtpa (5.4 bcm/y). In May 2012, GDF SUEZ signed an agreement with Cameron LNG to negotiate a 20-year liquefaction contract for 4 mtpa (5.4 bcm/y). The completed liquefaction facility is expected to be comprised of three liquefaction trains with a total export capability of 12 mtpa (16.2 bcm/y) of LNG. The liquefaction facility will utilize Cameron LNG's existing facilities, including two marine berths and three LNG storage tanks. The anticipated incremental investment is estimated to be $6bn. Cameron LNG expects to receive the required permits from the DoE and the FERC and enter into a turnkey contract in 2013e for engineering and construction services for the project. This shows that major downstream market players (especially Japanese buyers) are more and more willing to access directly US LNG.

As of May 2012, several projects with a total capacity of 102 mtpa have filed applications with the US DoE seeking authorisation to export LNG. If all these projects were approved and built, the US would become the number one LNG producer, far ahead of Qatar (current number one with 77 mtpa (104 bcm/y))!

Federal law gives US DoE the authority to revisit liquefied natural gas export applications it has approved. We believe this is unlikely as:

• Cheniere's Henry Hub linked formula will not make the US gas market oil-indexed dependent.

• A claw-back would have to mitigate a very serious threat where gas would not be available even for US citizens, and in this case HH would have gone up so much that exports would be uneconomical anyway!

• The US is already a net gas pipe exporter to Mexico (14.1 bcm in 2011).

– Reserves estimates are... uncertain!

In 2012, the US DoE reduced the estimated unproved technically recoverable resource of shale gas for the US to 14 tcm, substantially below the 2011 estimate of 24 tcm. The decline largely reflects a decrease in the estimate for the Marcellus shale, from 12 tcm to 4 tcm. Drilling in the Marcellus accelerated rapidly in 2010 and 2011, so that there is far more information available. But what matters, at the end, is production not resources. And as the DoE was reducing shale gas resources it was increasing shale gas production, that should represent 49% of US gas production in 2035e vs. 46% in the 2010 outlook (an increase from 336 bcm to 380 bcm (+13%) for 2035e shale gas production)...

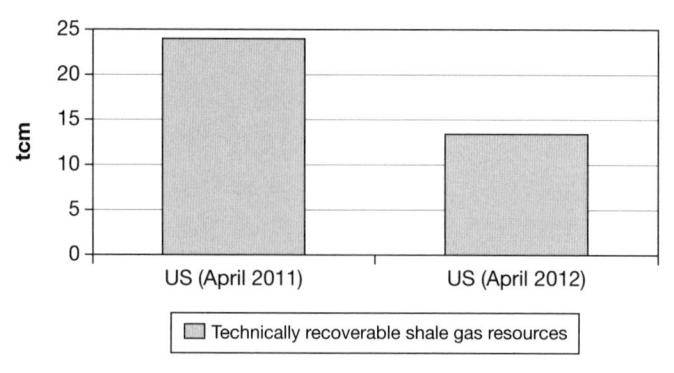

Figure 89

Estimated shale resources in the US.
Source: US DoE.

– *Canada needs to diversify away from the US*

The US is the only export market for Canada. Canada exports peaked in 2001; they are negatively affected as US need less imports thanks to its growing domestic shale production.

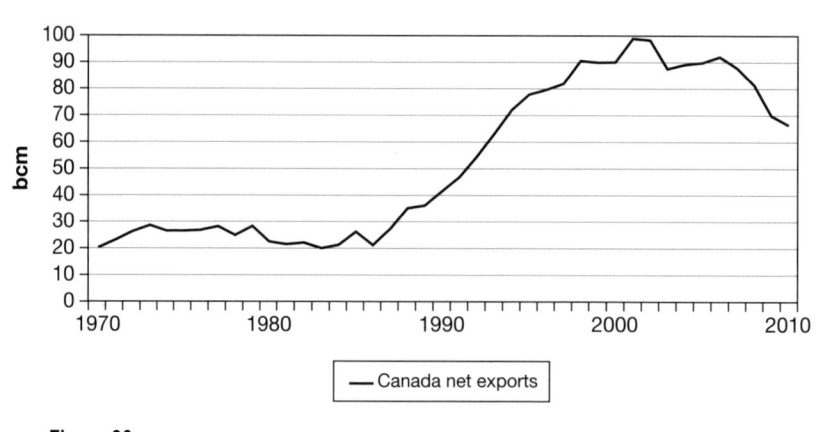

Figure 90

Canada net exports.
Source: BP Statistical Review.

This means that Canada, today 4th export country after Russia, Norway and Qatar, will need to find other markets by building LNG liquefaction trains to mitigate this drop in gas exports that should otherwise continue, if the US doesn't manage to boost its own LNG exports. This explains why already two Canadian projects, in Kitimat, have been granted by Canada's National Energy Board, a licence to export, in total, 12 mtpa.

CHINA HOLDS THE LARGEST UNCONVENTIONAL GAS RESERVES

– Outside the Americas, we believe that China could probably use its vast unconventional resources.

In April 2011, the US DoE published a study on shale gas outside the US (excluding Middle East and Russia), which showed that China holds the largest resources (1.5x more than the US alone).

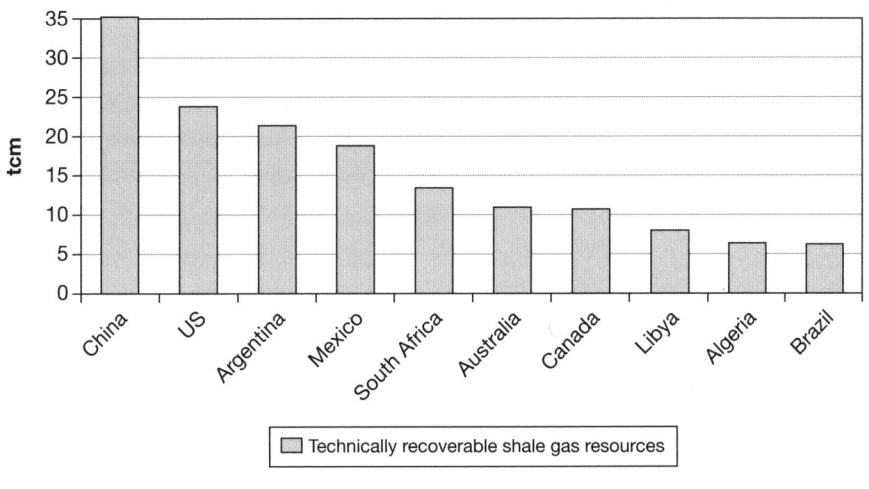

Figure 91

Major shale gas holders.
Source: US DoE 2011.

According to the US DoE, China is not only the world's major shale gas holder, but the country's technically recoverable shale resources are 12x its existing proved reserves (vs. only 3x for the US).

So if China was to follow the US model, its long-term shale gas production could reach 500 bcm/y.

In March 2012, the Chinese Ministry of Land and Resources announced that according to its survey, China onshore exploitable shale gas reserves are 25 tcm. Although the Chinese figure is lower than the US DoE one, it confirms that China is

Figure 92

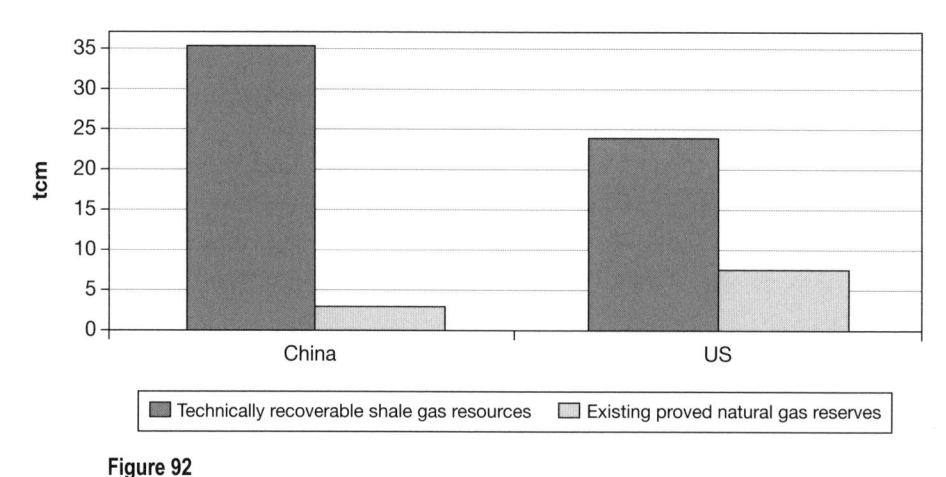

Shale gas resources vs. proven gas reserves.
Source: SG Cross Asset Research, US DoE 2011.

the largest shale gas reserves in the world. As we will see later, with Poland case, shale reserves are just estimates that need to be checked at each field level by drilling.

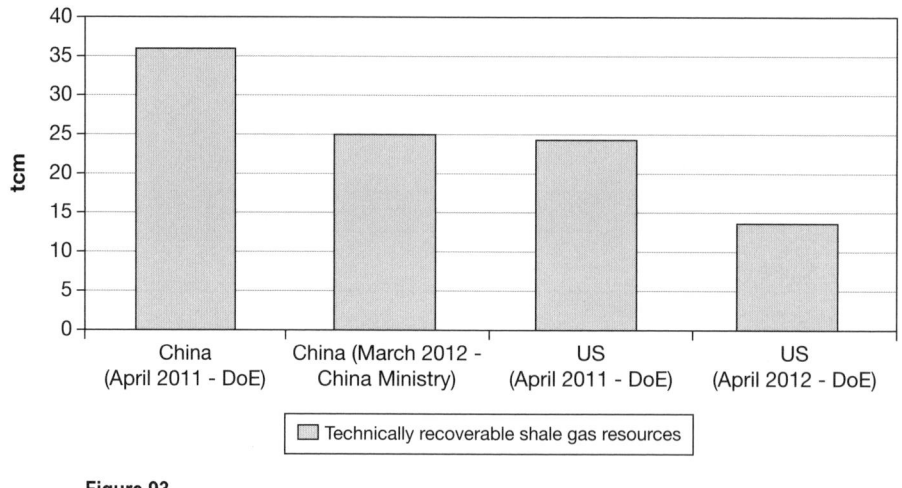

Figure 93

Shale gas recoverable resources are just estimates.

PetroChina began shale gas exploration, in December 2009, in the south-western Sichuan province but China needs the American technology to crack the problem and frack the reservoirs!

• China is accessing American Shale Gas Technology...

CNOOC paid $1.1bn in November 2010 for a one-third stake in Chesapeake's holdings in the Eagle Ford shale in south Texas (US). In addition, CNOOC has agreed to fund 75% of Chesapeake's share of drilling and completion costs up to $1.1bn, which Chesapeake expects to occur by year-end 2012.

In January 2011, CNOOC agreed to purchase a 33.3% interest in Chesapeake's oil and natural gas leasehold acres in northeast Colorado and southeast Wyoming. The project highlights that it is in the joint interests of energy companies in both the US and China to accelerate the development of shale oil and gas and increase energy supply.

In October 2011, Sinopec purchased Daylight, a company focused on exploiting numerous hydrocarbons plays in Alberta and north east British Columbia (Canada). Advances in horizontal drilling and multi-stage fracturing have had a major impact on Daylight's ability to more efficiently access the large natural gas deposits that have been recognised in this area.

In January 2012, Sinopec stroked a $2.2bn deal with Devon for a one-third stake in five US shale oil and gas fields, in another move by a Chinese state-owned firm eager to play a bigger role in the global rush, to tap unconventional fossil fuels.

And this "shopping list" should continue! But we believe China is also using IOCs to transfer US shale gas production technology (horizontal drilling + hydraulic fracturing) to China and implement it there. IOCs have very limited access to conventional gas in China, so obtaining interests in unconventional gas projects could be an additional incentive for IOCs to enter China, since the opening of the upstream sector to greater foreign participation in January 2012. European IOCs (which missed the US shale gas revolution and have little hope of using fracturing in Europe) would be more than happy to participate in China. On the other hand, China continues to adapt its legislative framework to make shale gas production a priority of the 12th five year plan development.

In March 2012, Shell signed a production sharing contract with China National Petroleum Corporation (CNPC) for developing a shale gas block in south-western Sichuan province, the first such deal in China.

An even faster way to boost unconventional gas production in China would be for North American independent gas producers to enter China with their technologies. But we have discounted this as both parties (independent producers and the Chinese government) would have major difficulties in adapting to the situation.

• ... and improving its Gas Infrastructure

China has the resources, the engineers and the manpower. IOCs and service companies could provide the technologies and the rigs to fast track the process of unlocking this massive resource. The main difference between the US and China remains the gas infrastructure. The availability of ample pipelines in the US has facilitated fast monetisation of shale gas. In China, by contrast, the infrastructure is obviously very limited. For the IOCs to commercialise the shale gas resource, the network of pipelines will need to be enlarged. As these pipes are built by state firms, the question is: would domestic gas prices need to increase to provide incentives for the state firms to build additional pipelines? Could that pose a threat to the scenario? However, as in the US, unconventional gas could prove cheaper to produce (and far cheaper than oil-linked prices), and the Chinese government might not need to raise regulated gas prices to spur this shift in production (the split between the commodity cost and the infrastructure could be altered to take into account the cheaper commodity cost and higher infrastructure cost).

• Production set to Grow by 13% CAGR until 2020e

On the production side, China has already seen high growth: the CAGR for 2000-2010 was 13.4% (CAGR for 2005-2010: 14.1%). Thanks to its new strategy to access unconventional technology, China is seeking to monetise its unconventional gas resources rapidly. So high production growth (2010-2020e CAGR: 13%) should continue and come from both conventional and shale gas production.

EUROPE: INCREMENTAL SUPPLY IN POLAND

– Environmental concerns have stymied shale gas operations in Europe, except in Poland which is keen to escape its reliance on Russian gas...

With high reliance on Russian gas (10 bcm out of 14 bcm demand) and potentially one of the largest unconventional gas resources in Europe, Poland decided to favour extraction of unconventional gas to reduce the country's dependence on Russia.

According to the US DoE (April 2011), Poland has the largest estimated recoverable shale gas reserves in Europe. But Poland's recoverable shale gas reserves could be lower than the US DoE estimates (5.2 tcm) as, in March 2012, the Polish Geological Institute estimated the shale gas resources to be between 346 bcm and 1.9 tcm. Both numbers are still estimates and more drilling is required to have a better view...

So when we compare US DoE data with new Polish data, we end up with France having the largest estimated recoverable shale gas reserves in Europe. This shows that DoE shale estimates can only be validated after effective drilling...

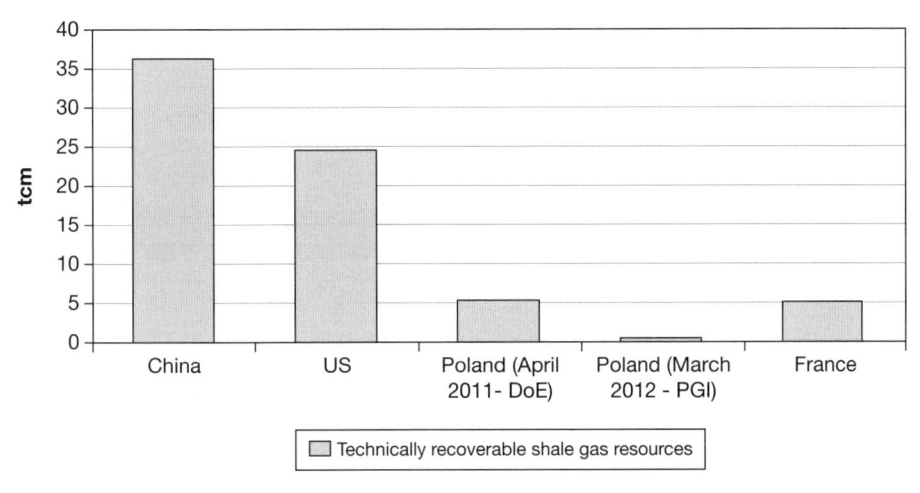

Figure 94

Poland or France, first in Europe?
Source: US DoE 2011 and PGI.

– *... but where are the Polish "sweet spots"?*

So far test wells in Poland have not managed to find the "sweet spots" (specific areas in the shale formation where production flows are much higher). The gas discovered in ExxonMobil's first wells in Poland, in January 2012, failed to flow in sufficient quantities to justify bringing them into production. ExxonMobil's failures followed disappointing results at Polish wells drilled in 2011 by other E&P companies. This supports our caution on achieving material near-term volumes as the industry needs to first find the "sweet spots". We believe that shale gas in Poland will start once the industry has discovered where the "sweet spots" are.

– *Can Polish production be economically viable?*

On top, the incentivised Russian take-or-pay contract signed, in October 2010, between Gazprom and the Polish company PGNiG (that is now in arbitration) was, we believe, an astute way for Russia to limit a future boom in supply from Poland. Incentivised take-or-pay allows the producer to sell extra volumes at a price that can compete with the cost of unconventional gas production.

Figure 95

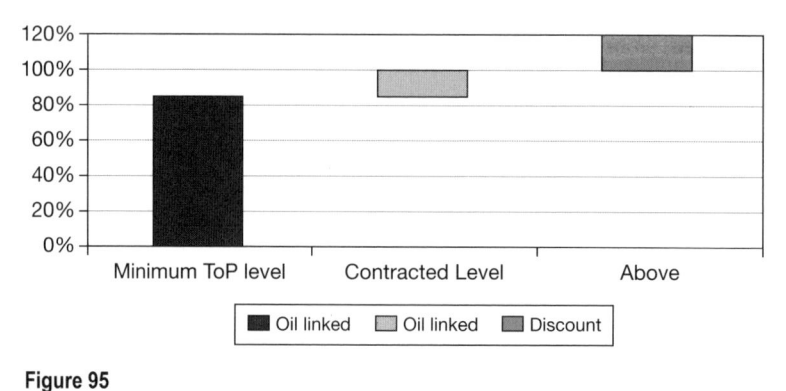

Incentivised take-or-pay allows the producer to sell extra volume at a discount.
Source: SG Cross Asset Research.

This leads us then to the cost of production of unconventional gas in Poland, where the geology is not as good as in the US and where the EU ETS means that the cost of CO_2 (when producing unconventional gas) has to be paid for in Europe from 2013e onwards. The cost of unconventional gas production could be around 4 $/MBtu in the US and around 9 $/MBtu in Poland; i.e. below current oil-indexation (13 $/MBtu). But Gazprom can still sell its conventional gas profitably even at this level, making it risky for producers to compete in Poland. Once "sweet spots" are found in Poland there will be consolidation as we've seen in North America and as we are witnessing in Australia.

Explorers in Poland confront rising taxes, a lack of rigs and rocks that are harder to drill than expected and potential government levy on production. ExxonMobil, Chevron and ConocoPhillips acquired rights in Poland, anticipating Eastern Europe's largest economy would lead the development of shale fields similar to those that upended North America's energy industry and made the US the world's biggest natural gas producer. Europe's greater population density and stronger environmental lobby make drilling more difficult than in the US. There are many differences between Poland and the US in terms of geology. Technology will have to be adjusted and costs will be higher than in the US. But Poland is now in a position to lead the shale industry development in Europe.

The minimum European price for Gazprom (with zero margin) could be 160 $/1,000 cm (4.5 $/MBtu). This shows that Gazprom has, thanks to this incentivised ToP contract, room to reduce its price for additional quantities to compete against any Polish shale gas production.

With the Russian mineral extraction tax flagged to increase substantially (up to 28 $/1,000 cm in 2015 so far) and cost escalation of new projects, we can estimate the total cost of production for Gazprom for 2020e to be around 56 $/1,000 cm (1.6 $/MBtu).

Figure 96

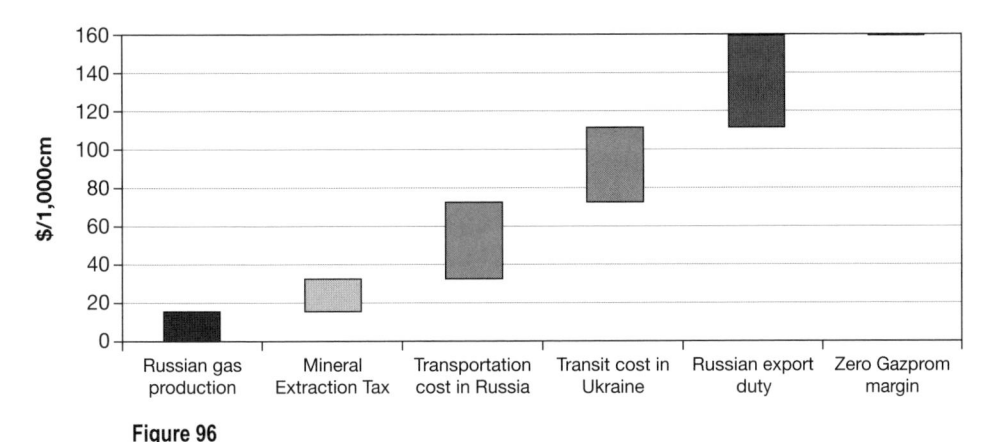

Minimum European price for Gazprom in 2012: 160 $/1,000 cm.

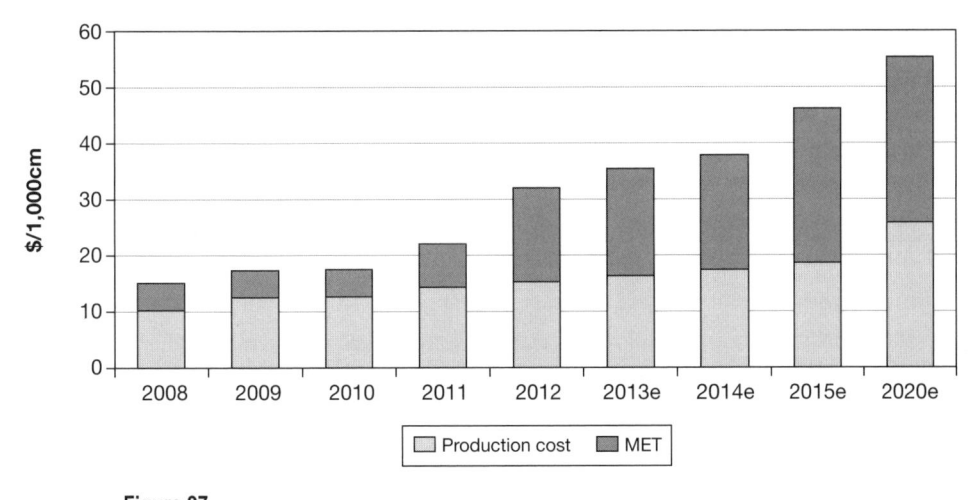

Figure 97

Russian cost of production and Mineral extraction tax for Gazprom's gas.

And the minimum European price (with zero margin) for Gazprom for 2020e could be around 210 $/1,000 cm (5.9 $/MBtu).

It is worth noting that Poland is pursuing an additional option to decrease its Russian gas dependency: LNG imports. Qatargas will supply 1.3 bcm/y of LNG to PGNiG under a 20-year long-term agreement, to start in 2014e. The LNG will be delivered into the new regas terminal (5 bcm/y capacity) in construction on the Baltic coast. This capex program seems to prove that Poland, as a country, thinks that it

Figure 98

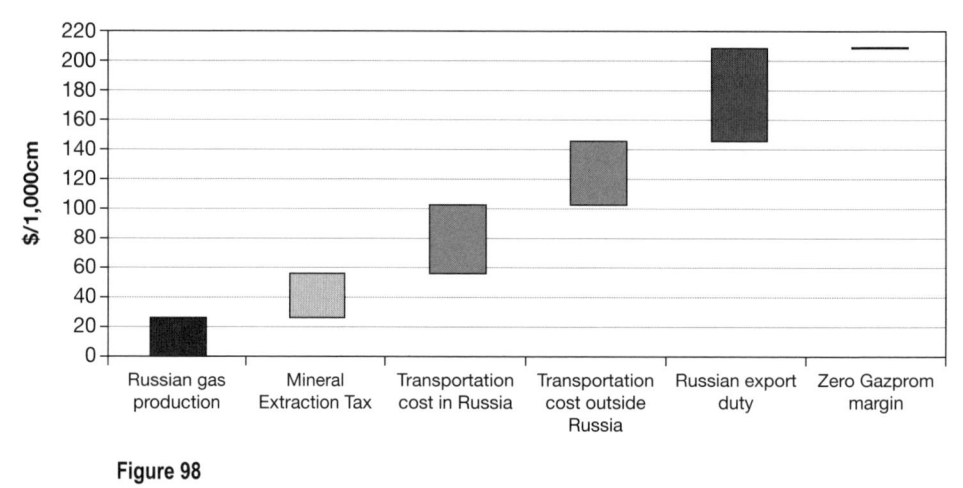

Minimum European price for Gazprom in 2020e: 210 $/1,000 cm.

will continue to be a net gas importer (and importing Russian pipe and Qatari LNG both oil-indexed!) even with its future domestic shale gas production, which is in line with our view.

• **Not in Continental Europe this Side of 2020e**

France (the biggest resource holder) and Bulgaria have banned hydraulic fracturing (respectively in 2011 and 2012) because of increasing fears over the environmental impact of hydraulic fracturing.

The Czech Ministry of Environment cancelled, in April 2012, an initial permit for exploration of shale gas deposits in the northeast of the country because the original permission had not fully followed the correct procedures with insufficient consultations with local councils and other interest groups taking place. A new application for a permit could be made but a decision would need to reflect public interest, protection of water resources, nature and the countryside. Legal changes might be necessary for the country to deal with a wave of shale gas applications.

In 2012e, Lithuania should launch an exploration license tender for shale deposits.

• **Perhaps Offshore in the North Sea**

Some unconventional production was tried in the UK but stopped due to tremors. Problems were raised in the UK over potential links to earthquake activity, as well as the chemicals included in fracking fluids. The UK government is still reviewing whether to allow fracking to go ahead again.

But offshore fracking technologies could mitigate the decline rate of conventional fields in the UK. As we've seen before, UK production decline has accelerated and reached 11.1% CAGR in the last 5 years. Even if onshore unconventional production is unlikely, companies could use fracking techniques for offshore fields. Drilling for oil and gas in waters off Britain increased in Q1 12 versus a year ago, providing a sign of what may prove to be an upward trend. A total of 11 exploration and appraisal wells were drilled in UK continental shelf in Q1 12, compared with 9 in the Q1 11.

This will need to be operated by focused companies. So first we should see the majors reselling their stakes in those fields to smaller companies that will then invest in those fields to stop the decline. Furthermore the UK government budget now provides greater certainty over tax relief for decommissioning North Sea production assets. The improved certainty on decommissioning liabilities would make it easier for smaller producers to take on assets from the majors.

The recent newsflow suggests this is starting in North Sea as unconventional offshore has a major advantage in Europe; it is offshore, reducing "Nimby" factor as fishes don't complain!

- In January 2012, Centrica has agreed to pay $223m for ConocoPhillips' 15% in the North Sea Statfjord oil and gas field. In February 2012, Centrica reached an agreement with Total to acquire their non-operated portfolio of producing oil and gas assets and associated infrastructure in the Central North Sea for a total cash consideration of $388m. The handling of those producing assets back to a UK dedicated company could be a new trend where majors leave this mature area allowing smaller companies to use new dedicated technologies to enhance recovery and mitigate the decline in UK gas production seen since 2000.

- In March 2012, BP has agreed is to sell its UK gas assets in the southern North Sea to privately held explorer Perenco for $400m.

- Wintershall started, in March 2012, production from its tight gas field in the Dutch North Sea. The well goes down vertically for approximately 3,750 meters and then has a horizontal section in the gas bearing sand of about 1,400 meters. Based on the good results of this first tight gas project, Wintershall will focus on the development of more tight gas fields in the Dutch offshore.

This could perhaps help stop the decline of the European gas proven reserves that we have witnessed in the last 10 years.

– For Europe as a whole, unconventional gas is likely to come only from Poland and should only amount to incremental supply, after 2016e, with a marginal price effect

While shale gas became a game-changer in the US, flooding the domestic market, in Europe, it could supply a useful diversification to boost energy security. With shale gas development in its early stages in Europe, the resource has the potential to play a marginal role in helping meet Europe's energy requirements this decade. The aim is to protect the environment while capturing the economic benefit.

European shale gas production could also be the only answer to the ill functioning EU gas market where 4 foreign NOCs control c.50% of the supply. Unfortunately the EU is pushing for a single energy market but not for domestic shale gas production.

– Watch out for the weakest link

As Poland will be the leading shale gas producer in Europe, the way the industry operates in that country will have a major impact for further shale (oil &) gas production throughout Europe. A tightly regulated Polish production process, with a systematic program for the disclosure of chemicals used in unconventional gas production, could help this industry expand in the rest of Europe. The move to reveal the make-up of fluids used in hydraulic fracturing of shale gas reserves (a similar initiative as the US fracfocus website (fracfocus.org) could be developed) would help to head off criticism of the new gas production method, which is attracting growing attention in European countries. A comprehensive disclosure program allows citizens and communities to consider this technology. Only this could lead to open discussion about environmental protection and risk management, and the potential benefits of shale development in Europe. Any environmental issue in Poland would have a dramatic effect on shale production throughout Europe. Poland, in the leading position, stands under scrutiny: if it succeeds, Europe will follow; if it fails, Europe will delay any shale production even further. The industry must understand that tighter environmental standards (and potential reduction in oil-linked prices) will mean that this business will not be as profitable as conventional gas production in major resource-holder countries… but the risks (financial, security, etc.) are much lower in Europe than in other gas producing countries.

Like any human activity, shale needs a "social license" to operate and the industry should be aware that its least successful player, in the eyes of the general public, defines the industry as a whole.

Shale gas production has transformed the US gas industry in recent years, boosting production rates and booked reserves of gas. The country was changed from a growing importer to a possible exporter, while US gas prices have dropped to 2 $/MBtu, against 9 $/MBtu in Europe.

It took 30 years of Research & Development in the US to unlock the shale gas resources. As understanding of unconventional resources improves, Europe could find a way to extract shale in a greener way (less water and air pollution) around the end of this decade. And even if costs would be higher than in the US, technology improvement could help to reduce those.

Development of shale resources gives rise to competing land use issues. The development and operation of shale projects requires a large number of wells, rigs and pipes. Projects will be competing for land, water and infrastructure with agricultural uses and communities. Skilled labour is also not in abundant supply and will need to get trained.

– Subsurface regime: not an issue

We don't think that the fact that in Europe, contrary to the US, the resources belongs to the State is an issue that could further delay shale gas production. First of all, even in the US, the resources are not exclusively in private hands. For example offshore resources (Gulf of Mexico) are federal resources. And even in Texas, the State has keept ownership of some acreages. Secondly, Europe is used to the system where states keep the ownership of subsurface, with the best example being the UK where the owner of a house is not the owner of the freehold… and must pay an annual ground rent. States are used to redistribute wealth and highly taxed shale production should help citizens to give the go ahead to local projects as long as local hospitals and schools are benefitting from those new revenues.

– Lack of service industry capability should be solved if and when "sweet spots" are found

The limited amount of land rigs available (11 in Poland in March 2012) shouldn't be viewed as a constraint. Once the "sweet spots" are found in Poland, the industry will be able to assess the maximum profitability of shale production and, if it makes economic sense, will need more rigs that the service companies would be more than happy to build and operate. As the exploration phase is just starting, Poland faces many unknowns ("sweet spots", public acceptability, regulatory regime, tax regime, profitability, ability of Gazprom to reduce its marginal gas price, etc.) that need to be solved before turning to a production phase.

AUSTRALIA COULD OVERTAKE QATAR IN LNG, THANKS TO UNCONVENTIONAL GAS

– LNG supply growth to resume in 2015e thanks to Australia

Thanks to unconventional gas, Australia is set to become the next growth area for LNG from 2015e. Australia's current 20 mtpa (27 bcm/y) capacity is set to grow, as a capacity of 57 mtpa (77 bcm/y) is already in construction and another

28 mtpa (38 bcm/y) could materialise before 2020e. This adds up to 107 mtpa (144 bcm/y) and could make Australia the number one LNG producer in 2020e.

Australia LNG has already been sold (before FID) mainly on an oil-indexation basis in Asia. So this extra gas shouldn't have an impact on future pricing.

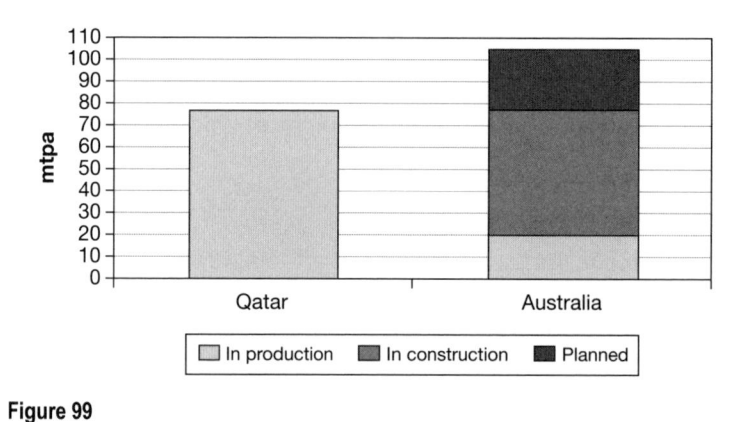

Figure 99

Qatar vs. Australia: LNG capacity.
Source: SG Cross Asset Research.

As Qatar, the lowest cost producer, was not prepared to compromise on a strategy of seeking prices close to crude oil parity, the highest cost producer, Australia, was able to stay in the competition and achieve prices that made the investment in its new projects economic. But the recent developments in the US and Canada could lead to North America becoming a major LNG exporter. US LNG should be much cheaper to build as:

1/ the upstream, transportation and LNG infrastructure (jetty, tanks) are already there;

2/ cost of labour is cheaper than in Australia;

3/ competition for water supplies (agriculture, industry and humans) is a major issue in Australia, water management of unconventional production is an ongoing and expensive operation.

Given their high capex requirements, Australian producers can only offer oil-linked LNG contracts… whereas Cheniere (and perhaps other US projects) are selling (and could sell) LNG under a Henry Hub linked formula. We therefore believe that US LNG supply could grow quickly in the 2016-2020e period.

In May 2012, BG announced a 36% rise in the capex of its Queensland Curtis LNG project since FID (half due to increase in local inflation, the rest from an A$/US$ appreciation). This announcement was followed by a 16% increase in

Figure 100

Disclosed capex of LNG projects.

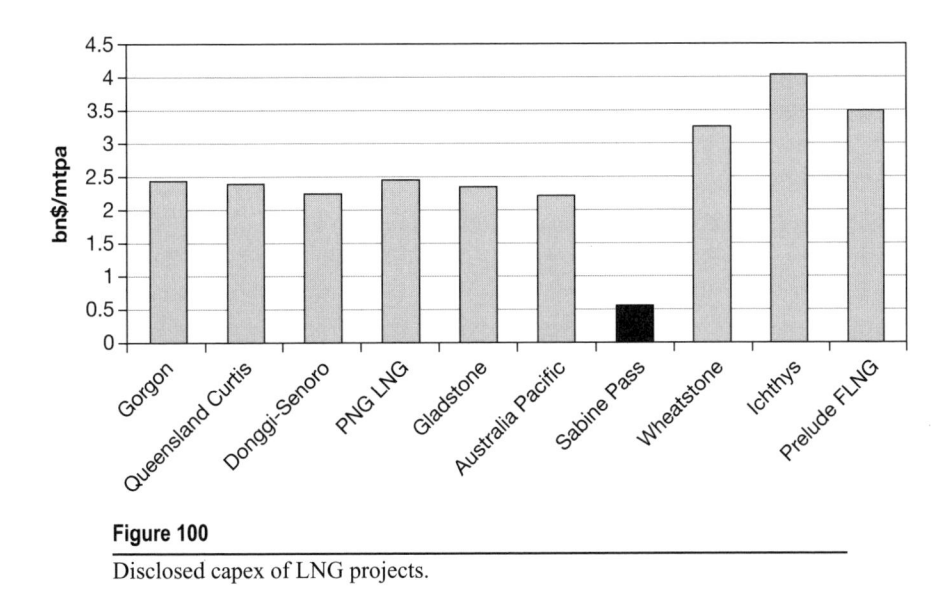

Gladstone's capex in June 2012 and could be followed by other Australian projects capex increases.

Ichthys could be the last green-field LNG project sanctioned in Australia because with rampant cost inflation in the face of an increasingly price-sensitive customer base, these large-scale, expensive projects simply look cumbersome and out-dated in the context of intensifying global competition. As a result, Australian projects are being priced out of the market. This coupled with delays is eroding returns from the country's already marginal developments. In the last two years, Qatar's pricing policy has meant that the highest cost producer, Australia, has been able to undercut the lowest cost producer, Qatar. The emergence of the US and Canada as potentially major LNG exporters will create a new environment in which Australia will find it more difficult to compete.

RUSSIA & NORWAY: A LITTLE MORE PIPE GAS BEFORE THE ARCTIC OPENING

• Russia: Major Projects under way could Face Challenges

Assuming it can reach its maximum daily output all year long, Gazprom would have been able to produce an extra 70 bcm in 2011 (but mainly in summer). But as pressure decreases in Gazprom's existing wells, the development of Bovanenkovo is critical to the fortunes of Gazprom, although the price tag for its development allows Gazprom little margin for error. At the same time, Gazprom faces challenges that threaten its dominance of the world's gas market: the emergence of US shale gas and the rise of LNG are transforming the global market, providing alternatives to Russian supply.

Russian investment in its gas production is increasing which will lead to higher gas availability, but, the gas is of higher cost. Russia will remain the largest gas supplier of Europe, but if it has the role of "supplier of last resort" then who will pay for insurance? It won't be cheap and Russia will definitely not pay. Someone has to pay for the spare capacity that will be idle in periods of low demand. Saudi Arabia can carry this role in oil as it has low costs; but Russia's cost of gas is rising.

The Yamal Peninsula is a region of Gazprom's strategic interests. 32 fields were discovered in the Yamal Peninsula and its offshore areas. The Bovanenkovo field is the most significant one. Projected gas production from the Bovanenkovo field is estimated at 115 bcm/y. In Q4 12e, the first gas produced from the Bovanenkovo field will be delivered to customers.

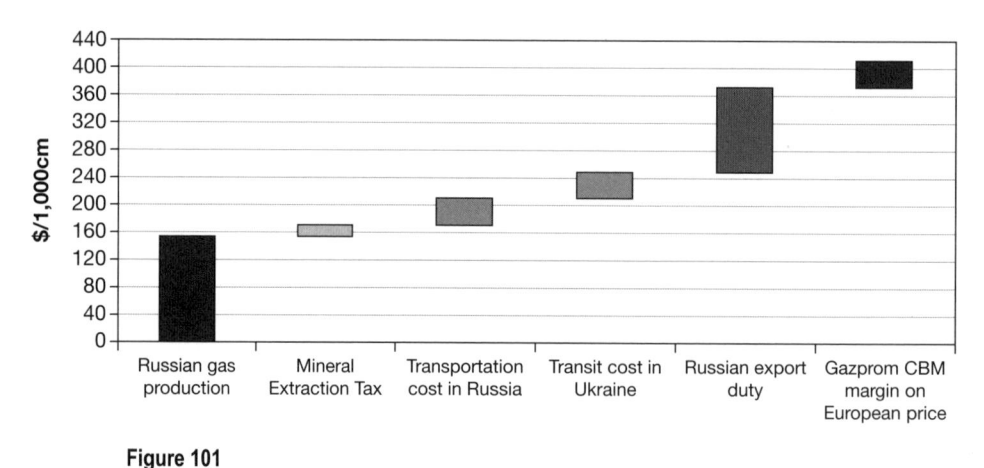

Figure 101

CBM in Russia, just profitable for Europe, not for domestic market.

In February 2010, Gazprom launched some unconventional gas production in Russia out of Coal Bed Methane (CBM) in the Kuzbass region. During the plateau period, annual CBM production in Kuzbass is planned to reach 4 bcm/y. We estimate the cost of this production to be 10x more vs. the traditional conventional Gazprom's production. At 155 $/1,000 cm, CBM can't be profitable in Russia (price of 90 $/1,000 cm) but it could be profitably sold to Europe at prices above 360 $/1,000 cm. Undoubtedly technology improvements will allow production cost to go South, and CBM to be even slightly profitable in Russia. But it is unlikely that we will see in Russia a US like situation where unconventional is cheaper to produce than conventional.

• Norway has the Flexibility to wield a Degree of Market Power but little Growth left in Traditional Areas

Norway wields a degree of market power in the UK, through its ability to arbitrage between the UK and Continental Europe. Its export network can handle 142 bcm/y, whereas its actual exports were 101 bcm in 2011 and should reach 120 bcm in 2020e. Norwegian gas is mainly sold under oil-linked long-term contracts in continental Europe, in contrast to which it sells on a spot NBP formula in the UK. But ultimately a seller will always try to maximise its income and Norway optimises the uses of its export pipes, particularly after the shock of the fall to negative prices, in the UK, in October 2006 during the testing of the southern leg of the Langeled pipe. Norway will not oversupply gas.

• Arctic Circle: Barents Sea first

Hydrocarbon resources under the Arctic Circle could be the next big race for, as global warming now allows to use the Northern route in summer from Russia to China (from Yamal to Asia the shipping time is reduced from 45 to 25 days, thanks to the Northern route). The first development could be in the Barents Sea as LNG could be shipped to European markets all year long. And Russia and Norway have finally sealed, in 2011, an agreement dividing a long-disputed area in the Barents Sea after decades of negotiations. The deal, which splits a 175,000 km^2 area into two equally sized pieces, has been approved by both countries' parliaments. This agreement opens up a promising oil and gas region in the Arctic, which has become more accessible through global warming. Exploration in the Barents Sea could prove fruitful but production shouldn't be seen this side of 2020e. As for the famous Shtokman and Yamal projects, they could be delayed until 2025e... All Arctic projects will need very favourable tax regimes (tax breaks, low mineral extraction tax, reduced export duty, etc.) to be profitable. This will have first to be decided by the Russian State for a long period of time allowing investors to make long term plans without the risk of possible changes if the State believes market conditions make it expedient. Hence why we discounted any gas from the Arctic Circle (pipe and LNG) arriving before 2020e.

OTHER PLACES

– Let's go briefly round the world to check what could happen (or not) before 2020e...

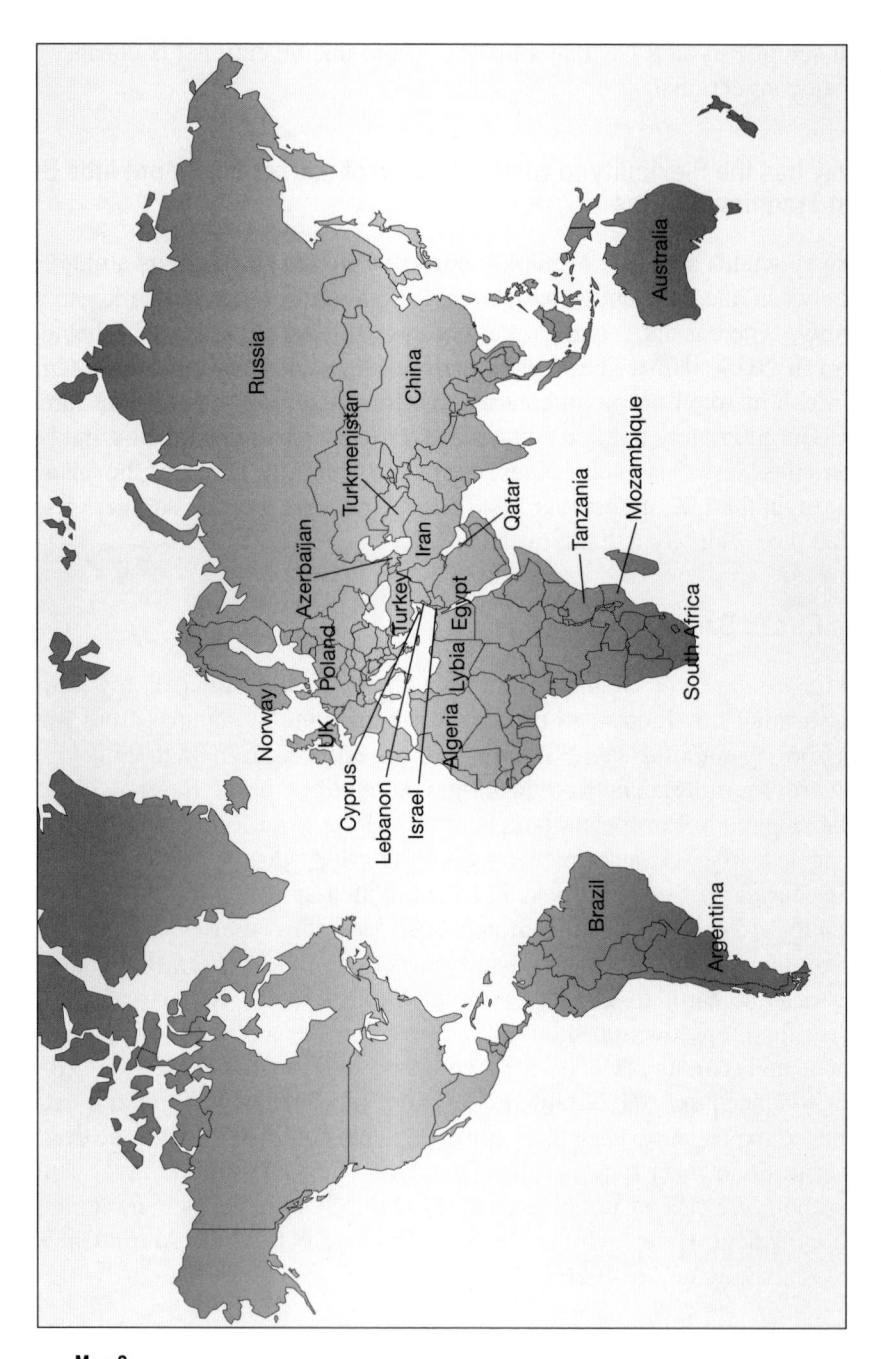

Map 2

The future supply growth?

- **South America**

  - *Brazil: offshore conventional for the domestic market*

  Major hydrocarbon discoveries have been done offshore Brazil since 2006. In this area oil production is growing fast, thanks to a first permanent floating production, storage and offloading (FPSO) in production since October 2010 (100 kb/d of oil and 1.7 bcm/y of gas). A further 12 FPSOs due to come on-stream progressively over the period to 2017e, should yield a total scheduled production capacity of approximately 2.3 mboe/d. To evacuate the gas across the Santos basin, BG and its partners are considering different options. In 2010, a pipeline was installed connecting onshore the Lula field. This pipeline has the capacity for three FPSOs (5.1 bcm/y) and further pipelines to other areas are under study, the first being planned for start-up in 2014e. The extra gas (20 bcm/y in 2020e) is likely not to be enough for the domestic market that is growing so fast that Brazil (+10.9% CAGR for 2000-2010) needs to import LNG since 2008. In 2010, Brazil demand was 26.5 bcm with 12.6 bcm imported.

  - *Argentina: shale perhaps but irrelevant for global markets*

  Argentina is the third major holder of unconventional gas resources but, because of booming demand, needs to imports LNG (since 2008). Low regulated prices, that have boosted demand, are a deterrent for upstream investment.

  Shale gas reservoirs, in Argentina, are deeper and less rich than in the US, except for the Vaca Muerte (Dead Cow) basin. On the positive side, shale formations are in areas of current conventional production and could benefit from the availability of surface infrastructure from declining conventional production. But government interference is high and increasing with licences being revoked and even the nationalisation of the Argentinian oil company YPF, formerly part of Repsol. This nationalisation could further delay unconventional production as it could be difficult for the new state owned company (that should get back the licences that had been formerly revoked from the Repsol owned YPF) to find the dedicated capex and/or other IOCs to invest.

  Even if Argentina (or Mexico) manages to produce shale gas, this shouldn't have a geopolitical impact as the HH is already a reference price not only in North America but also in South America.

- **South Africa: Shale stopped**

  South Africa government has halted shale licensing while it studies the impact of allowing companies to extract the fuel. Current regulatory regime is inappropriate and a new regulatory framework must be put in place that guarantees environmental integrity and equitable access to water. As this energy resource is critical for South Africa energy security, it is possible that once the new framework is decided, the residents could review their opposition as this resource is too valuable for South Africa to ignore. The Karoo shale gas could boost South Africa development but this can only be done once a full and effective new regulatory framework is set.

• Mediterranean Sea: Border Disputes likely to delay Israel LNG

Huge gas discoveries in the region from Israel to Cyprus have recently unleashed various plans for LNG exports. The eastern Mediterranean gas finds hold one of the world's biggest deepwater gas discoveries in the last decade, and more drilling is taking place:

- The Tamar field (250 bcm resources) was discovered in 2009. Noble Energy (36%) operates this field and should commission first gas at Tamar by the end of 2012e. Pipelines are currently under construction to bring Tamar gas to Israel in 2013e.

- Leviathan, with 480 bcm of mean resources, was discovered in 2010. Operator Noble Energy (40%) is evaluating the monetization options at Leviathan, including various LNG and pipeline export opportunities.

- In late 2011, Noble Energy (operator with 70% interest) announced a discovery offshore Cyprus and close to neighbouring Israel) with estimated gross mean resources of 200 bcm.

Tamar and Leviathan discoveries will first and foremost be used to meet the growing needs of the domestic market, notably in power generation. The Israeli government is reviewing policies to encourage fuel switching in the power sector with the goal of gas making 75% of the power sector supply by 2020. Until now, Israel was producing a small of amount of domestic gas and was counting on Egyptian pipeline gas but it has proven to be an unreliable source due to tensions between both countries, disagreements on pricing, and sabotage which has disrupted supply.

The maritime border issue between Israel and Lebanon could delay further resource development in Eastern Mediterranean Sea. While Israel asserts full ownership of the fields, Israel and Lebanon will eventually have to agree on a demarcation line by negotiation as other prospective fields in the region may be within Lebanese territory. And Turkey, which does not recognize the Republic of Cyprus, could make it difficult to monetise the resources offshore Cyprus…

Another key challenge for Israeli LNG will be the geopolitical risks associated to the overall instability of the region which could postpone the involvement of large international oil and gas companies. Those recent offshore discoveries should mitigate Israel growing demand and not be monetised as LNG on the global market this side of 2020e.

- **Caspian Sea**

  – Turkmenistan: for China

  With 12% of the proved reserves (end 2011), Turkmenistan has the reserves (and the Chinese money) to grow its production from its peak of 66 bcm in 2008, when it was exporting gas to Russia (and not China). In 2009, Turkmen production dropped to 36 bcm as Russia was not willing to take its gas as European demand was down. Since this low level, Turkmenistan has reoriented its exports East. And all the incremental Turkmen gas will now flow East as China has contracted 65 bcm/y (from 0 before 2010). So Turkmenistan could reach some 110 bcm in 2020e with all the increase dedicated to China.

  – Azerbaijan: for Europe

  In Azerbaijan, Shah Deniz (BP operator with 25.5%) was discovered in 1999. Shah Deniz Stage 1 started operations in 2006. The production from Stage 1 is 9 bcm/y and Stage 2 is expected to add a further 16 bcm/y. Stage 2 FID is expected to be taken in 2013, but it seems unlikely, as the export route has still not been selected, that this new gas could hit Europe before 2018e. Extra Azerbaijani gas at the end of the decade could materialise in Europe to help mitigate the decline of conventional European gas fields. It remains to be seen how this gas will flow into Europe, directly or via Russia as Gazprom is trying to secure long-term contracts with Azerbaijan to lock up those volumes. Gazprom buys actually 3 bcm/y from Azerbaijan but, as the pipe capacity is 7 bcm/y, this can easily be increased…

- **Offshore Africa: Unlikely even in Mozambique**

  New discoveries in Mozambique and Tanzania confirm that East Africa could become a key world gas supplier. As a growing number of companies, including Anadarko, ENI, Statoil and ExxonMobil, enter the race to lock up supplies, operators are making plans for LNG exports, with already two projects in Mozambique.

  Estimates for Mozambique's gas reserves keep being revised upwards and the country is now viewed as one of the most significant future gas player. Anadarko (36.5%) has increased its estimates of recoverable gas in its offshore Area 1 block up to 1,700 bcm. Anadarko plans to construct an LNG plant. Similarly, in addition to the Mamba South discovery in October 2011, ENI (operator with a 70% interest) announced in May 2012, that it made further discovery in the offshore Area 4, taking the total estimate of gas in place to 1,400 bcm. ENI also plans to export LNG. Given that ENI's and Anadarko's discoveries are technically adjoining blocks, building a common project would make commercial sense. And it is possible that the actual hyped news flow on growing resources by Anadarko and ENI is just a way for one of the 2 (presumably the one with the biggest resource) to take the lead in what could become a common project... Unless Shell that is taking over a small independent, in July 2012, (Cove energy that was a partner in Anadarko blocks) tries to further increase its stake there to take the lead as a major LNG player. But we doubt Mozambique LNG could materialise this side of 2020e. On top of this, it could make sense, to pipe some of this gas to South Africa (just on the other side of the border) that desperately needs energy and that can afford to pay for it! In 2010, Mozambique already exported 3 bcm to South Africa by pipe. But for those new projects to be piped to South Africa, a fair price will have to be agreed by all partners and both States... Mozambique has a favourable investment framework. With the prospect of a hydrocarbons bonanza, government will review existing oil and gas legislation and operators in Mozambique should anticipate tougher regulation and fiscal terms as the sector matures.

Tanzania, which is in a less advanced frontier exploration stage than Mozambique, shows promising prospects:

- Statoil and ExxonMobil, which are jointly developing offshore deepwater block 2 (65%), announced, in June 2012, that they have made gas discoveries so far proving up to 230 bcm of gas.

- BG (60%) announced, in May 2012, its fifth Tanzanian gas discovery in Block 1, offshore southern Tanzania. Prior to this latest discovery, BG had estimated mean total gross recoverable resources at 200 bcm of gas.

Those recent gas finds will apply pressure on the government to finalize its new draft legislation for the sector. The current Petroleum Act of 1980 covers oil and hydrocarbons, but not gas specifically, leaving the gas sector regulated by individual contracts, not a sustainable option given the growing interest and plans in the sector. The government is currently drafting a natural gas master plan to guide the exploration, production and transportation of gas as well as a petroleum revenue management bill. But again this will take time...

A final major challenge is cross-border collaboration between Mozambique and Tanzania. For now each company is focusing on its own discovery and export project to show up results to their shareholders. Absence of cooperation could represent a future stumbling block to the profitability or viability of projects.

East Africa is very well placed to serve the high growth Pacific basin market. Asian nations are in a hurry to secure as much supply as possible from existing and new suppliers. Japan, South Korea, but also China, India and South East Asian countries, notably Malaysia, have shown interest in potential LNG exports. But to gauge the competitiveness of East African LNG, the key questions are time and money. A growing number of LNG projects are competing for the same markets and window of opportunity. East African LNG will have to compete with existing Pacific suppliers and new emerging suppliers from potentially North America. Due to its Indian Ocean location, Mozambique and Tanzania will offer reduced shipping costs to India and Asian markets. But the money question remains and we doubt that in less than 10 years it is possible to produce LNG from scratch: upstream discovery, gas legislation, contracts negotiations, FID and finally LNG production, each step takes time. The lead time of this industry is decade and the financial unit is $bn…

• North Africa

Algeria (2.2%), Egypt (1.1%) and Libya (0.7%) together hold 4% of the worldwide proven gas reserves. The recent regimes changes in Libya and Egypt as a result of the "Arab Spring" pose a potential opportunity to increase oil & gas production. But booming domestic demand and internal politics have and could continue to hamper exports. As for Algeria, the production has been on average 82 bcm since 2000 and no significant growth is expected to materialise this decade. The reconstruction of Skikda and the construction of Gassi-Touil should, in the second part of this decade, help Algeria grow its LNG exports. As for Libya, the exports via the Greenstream pipe to Italy that have stopped during the civil war in 2011 have now recovered nearly to their former level but the Marsa el Brega LNG plant is still not producing any LNG.

• Middle East

With Iran and Qatar having 28% of the worldwide reserves it could make sense to expand production in those countries but, as we've discussed earlier, we believe Qatar's moratorium is there to stay until Iran is able to produce LNG…

– The wild card: a regime change in Iran

Iran's nuclear program has become the subject of contention with the Western world due to suspicions that Iran could divert the civilian nuclear technology to a weapons program. This has led to impose sanctions against Iran, thus furthering its economic isolation on the international scene. A regime change in Iran (as a result of any kind of action) could prompt a fast change from the West if the new regime decides to stop its nuclear program. With the second gas proven reserves on a worldwide level, a "new" Iran could benefit from IOCs (mainly European) and Japanese companies to invest heavily in liquefaction trains to monetise the South Pars field, the Qatari are already producing on their side.

Where is the Future Demand Growth?

New markets might be created by virtue of any carbon pricing mechanism, which should act to promote cleaner fuels such as gas as a transitional fuel until we are able to avoid or sequester all CO_2 emissions. But as we explained earlier this would need some policy will and tools and enough wealth for the citizens to be able to pay for CO_2 pollution. This could be one of the major wild cards in forecasting gas demand. We have made some hypothesis when going round the world on a regional level but depending on how citizens react to global warming those policies could be fast tracked or delayed...

As global warming is a major concern for customers as well as political leaders, gas is now a transitional fuel that has to compete against other energies.

IN ASIA, MAINLY IN CHINA

It is interesting to note that, for oil, China represents 11% of global demand (after US (22%) and EU (16%)), when for gas, China represents "only" 3% (after US (22%), EU (16%), Russia (13%) and Iran (4%)).

And India represents less than 2% of worldwide gas demand...

• China Consumption set to Grow by 15.5% CAGR out to 2020e...

China's 2020 strategic energy plan aims to increase gas in the country's energy mix from 4% to 10% in 2020e. Furthermore, nuclear new build in China have slowed down after the Japanese disaster. We have a Chinese gas consumption forecast of 15.5% CAGR for 2010-2020e (20% p.a. at the beginning of the period, slowing to 11% p.a. at the end).

In a decade (from 2010 to 2020e), China gas consumption should move from the level of Japan to the EU! In 2020e, for gas as well as for oil, China should be the third worldwide consumer after the US and the EU.

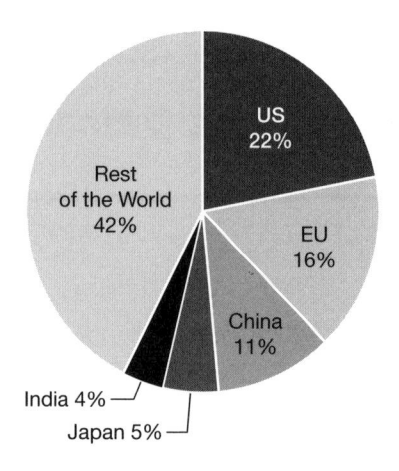

Figure 102

2010 split of oil consumption.
Source: BP Statistical Review.

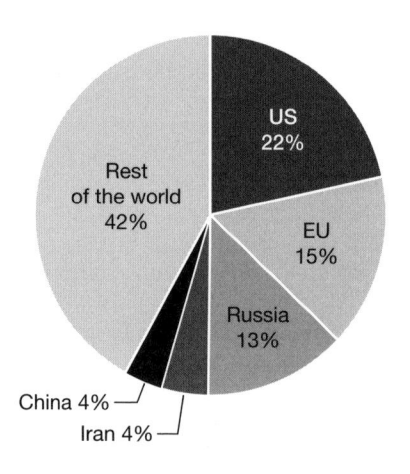

Figure 103

2010 split of gas consumption.
Source: BP Statistical Review.

• India

In the last decade, India, with a CAGR of 8.9%, has been the fourth fastest growing market after China (15.2%), Brazil (10.9%) and Turkey (10.4%). This trend should continue thanks to growing power generation. But the difficulty to forecast Indian gas demand is linked to the hick ups of their major gas field (KG-D6) and the Administrative Price Mechanism (4.2 $/MBtu since June 2010). The government

Figure 104

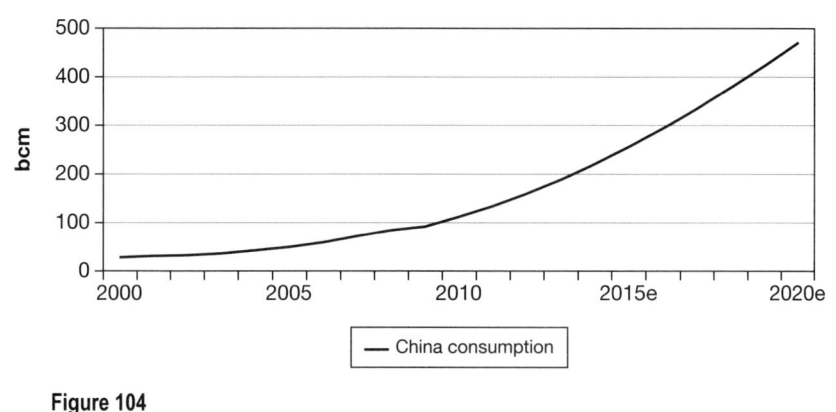

China consumption.
Source: SG Cross Asset Research, BP Statistical Review.

sets and manages pricing of gas for protected/priority industrial sector. But with such low prices, Indian companies can't buy LNG at international prices, leaving the country short of gas… Shale gas licenses should be awarded via competitive bidding by end 2013e. If this proves successful, there could be a race off between domestic shale gas and LNG, but the question of unconventional pricing remains.

• Japan: 2012e could be the Start of a Plateau Demand

In Japan, nuclear generators must be shut for inspection at least once every 13 months. The maintenance period can vary from a few months to more than a year, and the restart typically begins with a two-month test run before advancing to commercial operations, a step which requires regulatory approval, which was not given until July 2012. In May 2012, Japan had no nuclear reactor left operating (out of a total of 54). So, the diversion of LNG cargoes to Asia that reduced the UK's LNG imports in the second half of 2011, has accelerated. Since January 2012, Spain's LNG imports were once again higher than those of the UK. The growth in LNG demand in Japan (witnessed in 2011 and in 2012e) due to the nuclear shut down could reverse in 2013e if Japan decides to put many nuclear reactors back on-line in H2 2012e… We assume that Japan will put some nuclear reactors back to avoid burning oil but not enough to reverse the LNG spike witness in 2012e that we estimate at 123 bcm. From this peak, Japan gas demand will stay flat for the remainder of the decade due to energy efficiency.

An orderly nuclear phase out in other European countries (Germany, Switzerland) has and should not create extra gas demand (see Europe) as it should mainly be replaced by renewables. In North America, the cheap gas prices have derailed the nuclear renaissance even before the Fukushima disaster…

FINDING NEW DEMAND FOR GAS IN NORTH AMERICA

• To generate more Electricity

As seen earlier, gas is in competition with many other fuels to generate electricity. But cheap US gas prices have put a stop (even before the Fukushima disaster) to the US nuclear renaissance. At today's HH prices it is unlikely that a company will commit to a capex program to build new nuclear reactors in the US. With time, gas thanks to its low pricing should increase its stake in the electricity mix in the US.

• As a Transportation Fuel

The US is a mature gas market but cheap gas prices could boost new gas uses. On top of displacing coal in power generation it could be used as a transport fuel. But adapting the US transport system to new fuels (CNG and/or LNG) has always been viewed as the chicken and egg problem: what needs to be addressed first, the trucks and cars or the refuelling stations? "How can I guarantee a full tank over the US with my new gas car?" is the question from any potential car buyer when "How can I guarantee my refuelling station is going to be used if I'm a first mover?" is the question from the industry…

With gas prices so low it is the vested interest of US gas producers to find a new demand for their gas.

With less than 1% of Natural Gas Vehicles out of the total amount of vehicles on the road the US has an enormous room for growth. CNG vehicles and LNG vehicles are been put in service either for hard trucks uses but also for buses. Small and mid-scale liquefaction trains are being built. The use of LNG as an alternative to Diesel is being pushed by major US gas producers (Chesapeake and Encana) for their own fleet of vehicles. This should lead to the infrastructure to be developed and then people being offered the choice of a natural gas vehicle or a standard vehicle. The incentive would then be for people to adopt vehicles that run on the cheapest fuel, i.e. gas (on a Diesel Gallon Equivalent, CNG was priced at 2.29$, LNG at 2.92$ vs. diesel at 3.96$ and gasoline at 3.34$ (price end May 2012)).

Chesapeake plans to invest $1bn in the US over the next 10 years to stimulate market adoption of CNG, LNG and Gas To Liquids (GTL) fuels. Clean Energy Fuels wants to build America's natural gas highway allowing LNG truck fuelling coast-to-cost and border-to-boarder. Clean Energy Fuels build, operate and maintain fuelling stations that compress and dispense CNG fuel and dispense LNG fuel at strategic locations along major trucking corridors to form the backbone of a US national transportation fuelling infrastructure. The company fuels more than 6,000 natural gas buses across North America every day, supplying both CNG and LNG to transit fleets that include Los Angeles, and Vancouver (Canada). Today, 25% of all new transit buses on order are natural gas-powered.

US car manufacturers should soon release pickup trucks powered by natural gas. The biggest hurdle to wider use is refuelling. Today, there are fewer than 400 public CNG fuelling stations in the US.

Those news gas uses, as a transportation fuel, will compete against oil and can only take off if gas is priced differently than oil…In the XXth century, gas was priced vs. oil (with a small discount) to win market shares in domestic heating. In the XXIst century, gas could be used as a transportation fuel if it is priced much cheaper than oil! And today this oil vs. gas competition could even be easier (for gas) than the coal vs. gas competition for power generation.

• US Plastics Exports

Chevron Phillips Chemical is spending $5bn to build a new ethylene plant in Baytown, Texas, by 2017e as well as two polyethylene plants and related infrastructure. The industry may build US factories that convert natural gas into plastics because shale gas has made American production the cheapest outside the Middle East. Cheap gas is doubly advantageous to chemical makers because it's used as a raw material and to power factories. US plastics exports may surge as new plants start.

– More petrochemicals in the US means less somewhere else…
According to the DoE, petrochemical uses of gas were reduced in the US by 15 bcm since 2006 to 25 bcm today (out of a total industrial demand of 190 bcm). This industry could regain those losses… at the expense of Europe!

NOT IN EUROPE, UNLESS NEW USES ARE FOUND

According to the BP Statistical Review, UK gas demand has reached a plateau in the last 10 years, with 2010 consumption exactly at the level of 1999 (94 bcm), showing that the most mature gas market in Europe has been stagnant.

And growth in Spain is now behind us, as Spanish gas demand peaked in 2008, prior to the subprime crisis. Since then, demand has been falling…

Finally, in 2011, Europe saw a massive drop in gas consumption due mainly to a record warm year (2011) vs. a record cold 2010. Underlined consumption growth is going to be so low that Europe gas demand changes this decade will only be linked to weather. Therefore going forward, we believe European gas demand would stay in between the records seen in 2010 and 2011, with an average at 500 bcm.

Figure 105

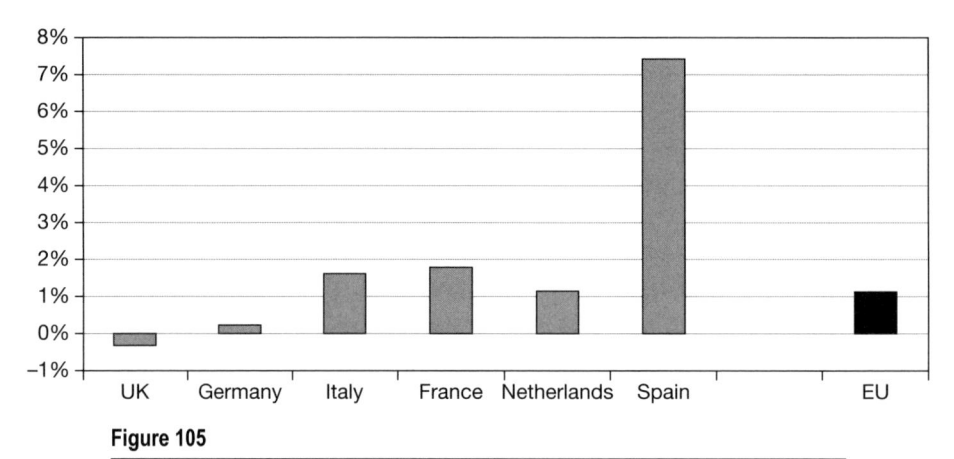

2000-2010 gas demand CAGR in six major EU countries and the EU.
Source: SG Cross Asset Research, BP Statistical Review.

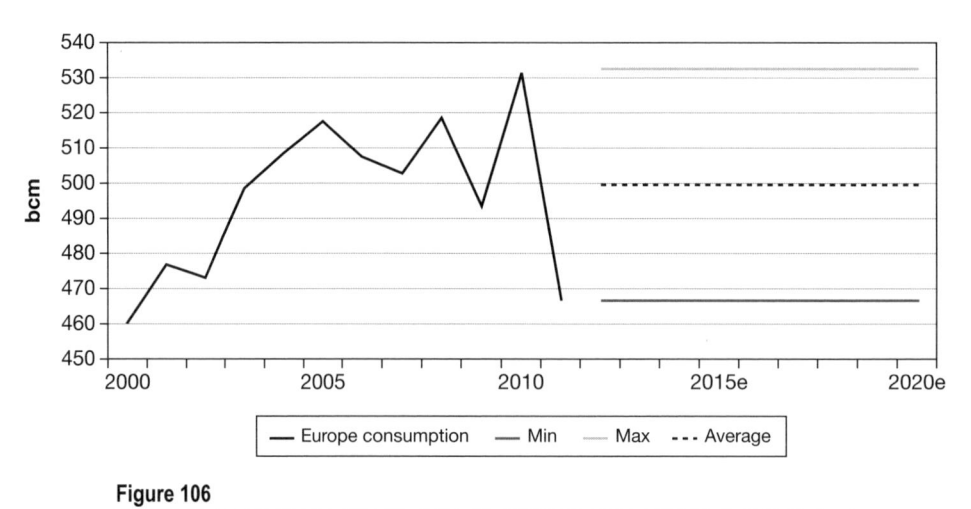

Figure 106

Europe gas demand.
Source: IEA for historical data.

The 63 bcm between the minimum and the maximum is therefore just linked to a weather play.

On top of this, as explained, a booming chemical industry in the US means a risk of closure of chemical plants in Europe, reducing further gas demand… So we could be, at the end of the decade at 485 bcm, on average.

Figure 107

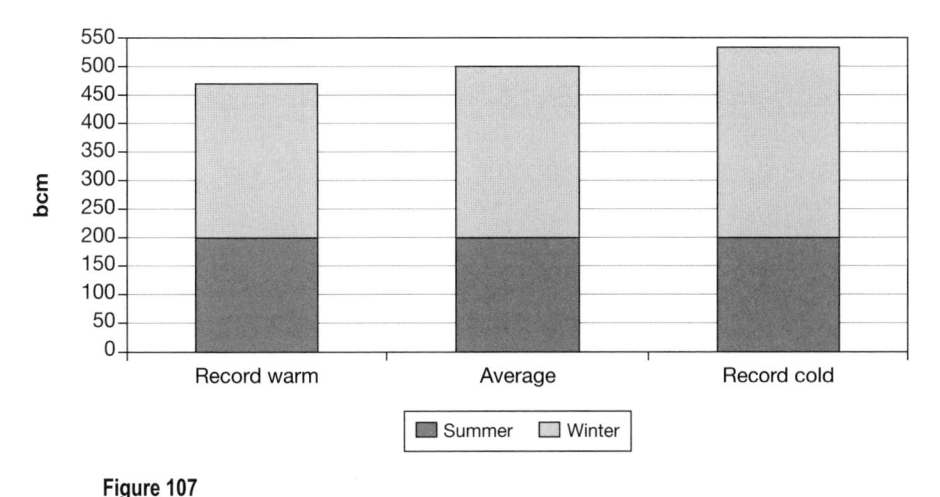

Split between winter and summer gas consumption in Europe.

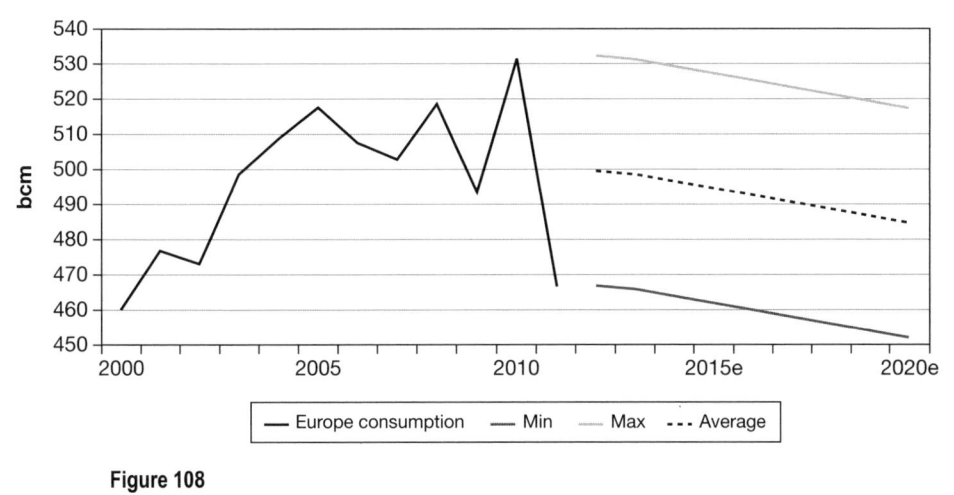

Figure 108

Europe gas demand assuming a 15 bcm loss from industrials.
Source: IEA for historical data.

Perhaps, once the EU energy market is completed in 2014 and once the European gas price discovery is based on transparent markets, the industry could be more willing to re-invest in gas in Europe. But right now, poor profitability (if any!) pushes European utilities to avoid any new investment in gas. As in the US, new uses could come from the transport sector.

• What about Global Warming?

Is global warming going to provide us with warmer winters or could we see warmer temperatures on average together with periods of severe cold? This unanswered question could have an impact on European gas demand.

• Gas to balance Renewable Energy Production

We expect a decline in residential consumption, thanks to improved building insulation. On the power side, even if we see an increase in power demand (linked to GDP growth), gas is only one of many fuels available. Gas is, in particular, in competition with renewables (depending on availability) and coal (depending on market prices and CO_2 costs). On top of this competition lies the question about nuclear, where phase out has been decided in Germany in Switzerland but where France is building a new plant. Gas is in fact the balancing fuel to take over when renewable energy production stops (no sun, no wind). Gas plants are the fastest backup sources of electricity when weather-dependent output falls. With more and more renewable entering the EU energy mix, gas consumption should plateau this decade.

• To be Green or Profitable?

– Decline in gas demand is also due to relative appeal of coal as a power generation source

As far as demand from the power sector is concerned, rising energy prices across the board imply that gas now has to compete and is no longer viewed as the "fuel of choice". Spark spreads indicate that coal-fired plants will be more economic to operate than their gas-fired counterparts.

As we've seen earlier, in the UK, since April 2011, it has been more profitable to burn coal than gas (priced on NBP). In Germany, gas has not been a profitable fuel for power generation since over a year. Fukushima pushed Germany to close all its nuclear reactors. But as coal has become a profitable fuel for power generation, nuclear (CO_2 free) has been replaced, in Germany, by the most polluting fossil fuel: coal! The rising cost of gas linked to oil price has wiped out the returns from burning gas. Germany will have to close old plants to counter new gas- and coal-fired stations that may come online in an already oversupplied market.

Most power markets in Europe have spreads that are signalling that today gas plants should be closed:

• In the UK, so far 4 uneconomical gas-fired power plants have closed in 2012 as companies are managing their fleet by taking offline the least efficient stations.

Figure 109

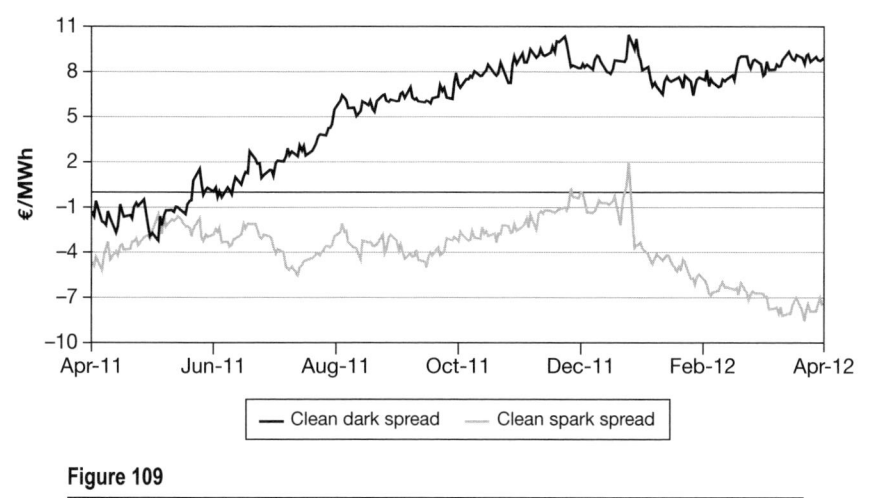

German Next Calendar Clean Spreads.
Source: SG Cross Asset Research, Reuters.

- The biggest losses since 2009 from burning natural gas to generate electricity in Germany are also threatening to provoke a wave of power-plant closures in Europe's biggest economy.

  – Gas demand in Europe has reached an undulating plateau and variations until 2020e will mainly depend on temperatures.

THE FUTURE OF OIL-LINKED CONTRACTS

– Demand growth is linked to price. It is therefore important to understand what is happening in Europe in terms of oil-linked pricing as this could have implications not only on demand in Europe but also on potential changes in Asian pricing as we will discuss later.

• Oil-linked vs. Spot

Long term oil-linked contracts are creating two problems for European customers:

- They are much higher than spot prices since the financial crisis of late 2008.
- They have minimum contractual take or pay quantities that push buyers to have to "pre-pay" for some gas that they can't take.

Gazprom has long-term contracts to sell above 150 bcm/y (154 bcm for 2012e) of gas to areas it classifies as 'Far Abroad' (Europe + Turkey). This means that, since 2009, Gazprom's buyers in these areas have been confronted by two issues: minimum take-or-pay (ToP) volumes and negative margins when reselling out-of-the-money, oil-indexed gas. When buyers take less than the minimum take-or-pay volumes, they have to 'pre–pay' the 'virtual' volumes not taken, with a roll-over period that can last up to the duration of the contract.

During Gazprom's Investor Day, in February 2012, Alexander Medvedev, Director General of Gazprom Exports, confirmed the existence of Gazprom's 'pre-paid' gas, amounting to 4 bcm in 2009, 8 bcm in 2010 and 1 bcm in 2011, totalling $4bn.

Alexander Medvedev explained that "Gazprom's long-term contracts portfolio ensures the sale of a minimum of 4,000 bcm to Europe over 2012-2030".

As Russian gas was more expensive than Norwegian gas from 2009 to 2011, European buyers concentrated their take or pay obligations with Gazprom. The company the most affected by those obligations is ENI with, so far, €1.5bn pre-paid. In 2011, ENI managed to reduce its ToP obligations thanks to the drop of Libyan gas. With the end of the civil war and the return of Libyan gas, ENI could still face some Russian ToP obligations in 2012e.

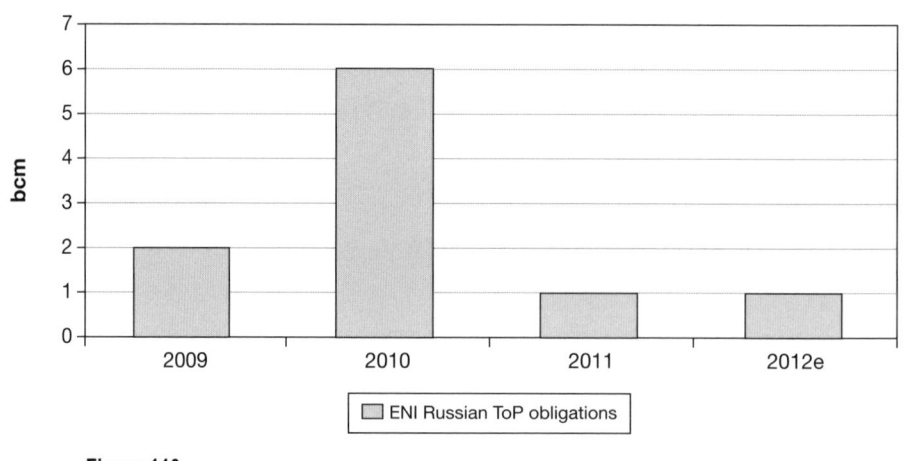

Figure 110

ENI Russian Take or Pay obligations.
Source: SG Cross Asset Research.

The oil-indexed price is based on an average of the previous six to nine month prices of fuel oil and gas oil.

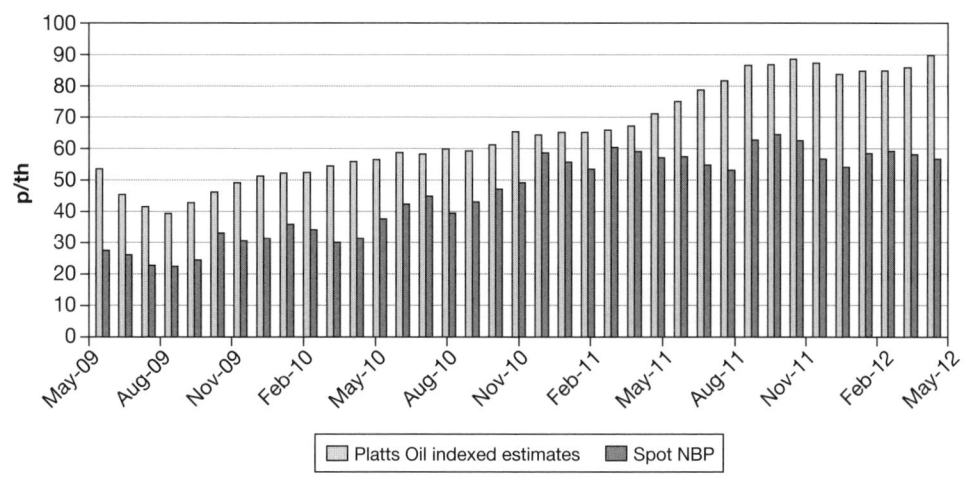

Figure 111

Oil index vs. spot prices.
Source: SG Cross Asset Research / Platts / Datastream.

The high spot/oil-linked price gap continues to put pressure on the major utilities, such as E.ON and RWE which import gas under oil-linked prices but are called upon by consumers to sell at spot-linked levels. By adapting production to demand, Norway makes sure this gap doesn't widen even further, thereby, de facto, helping European utilities not to report even higher losses, while increasing its own gas rent.

With power prices completely liberalised in Europe it now appears impossible, and unacceptable, for utilities to commit to further oil-linked long-term gas procurement contracts. This rules out the use of oil indexation contracts to purchase any extra gas needed to replace nuclear. It is therefore now time for major gas producers (notably Gazprom) to think about alternative indexation methods for pricing such additional gas requirement. **Either producers stick to oil indexation and there is a possibility that no new contract is signed because European utilities are unable to bear any additional risk, or they try to meet the demand for additional power generation and provide contracts acceptable to the power industry and we could witness a renewed dash for gas.**

During 2009-2011, big gas producers resisted major formula changed. They preferred to reduce their level of production while still enjoying high rents thanks to high oil-linked gas prices. But this strategy could help, in the long term, new competitors entering the European gas market. To avoid this risk, big gas producers are compelled to move, either by agreeing to alternative pricing (hub or electricity based) or by taking stakes in electricity producers.

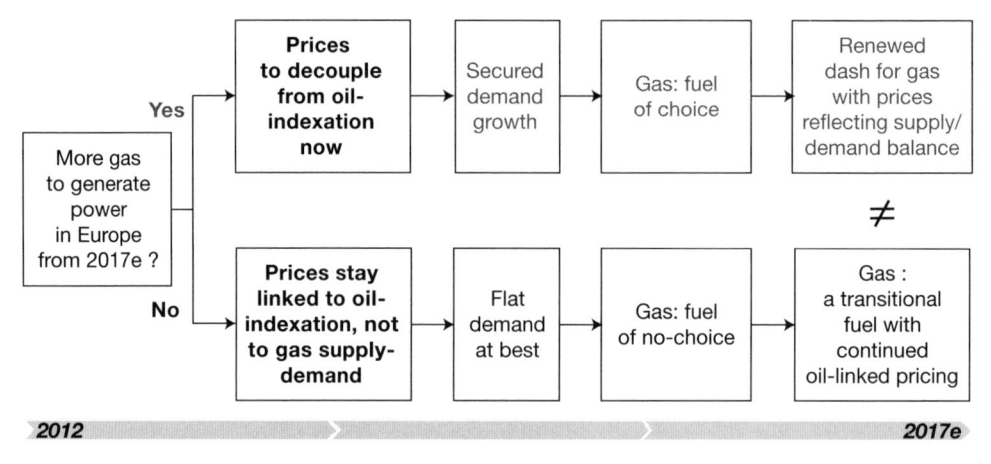

Figure 112

The conundrum: oil-linked prices or renewed dash for gas?
Source: SG Cross Asset Research.

Producers have to be innovative if they want to grasp the opportunity created by baseload electricity losses due to the nuclear phase-out. Even Gazprom recognised that the long-term contract formula needed to be addressed to enhance the competitiveness of Russian gas in Europe.

• For New Gas Volume to replace Nuclear, Gas has to be priced Differently

Low clean spark spreads put significant pressure on European utilities and the market could see further gas generation turned off in favour of coal. This suggests the oil price linkage needs to break to allow gas-on-gas competition and to enable those markets to function more efficiently.

If electricity is at 53 €/MWh and CO_2 at 7 €/t, gas needs to be priced below 25 €/MWh for a power plant to generate a positive margin… Far below the oil-indexed level of 37 €/MWh.

– Oil-linked price with a discount

In July 2011, Edison and Promgas (then a 50-50% JV between ENI and Gazprom and now 100% Gazprom owned) successfully completed renegotiations on the price review for the long-term contract for the 2 bcm/y supply of natural gas from Russia in order to comply with changed market conditions. Like ENI in 2010, Edison has finally achieved a discount on oil-indexation. Under the new agreement, from end of 2009 on, the gas price paid by Edison will be an estimated 10% lower price than in the previous contract.

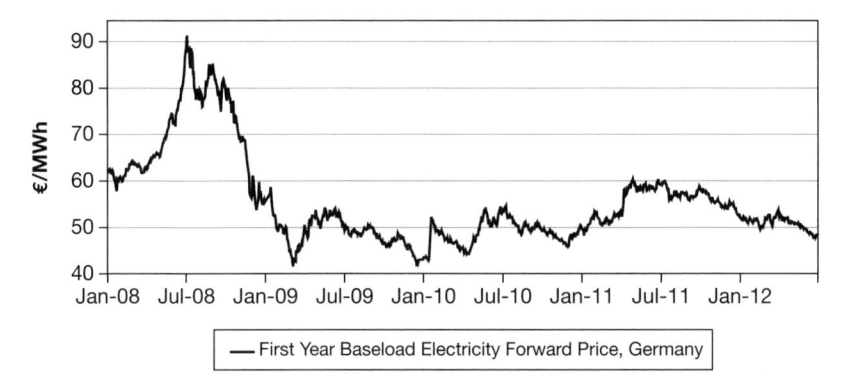

[Figure: line chart showing First Year Baseload Electricity Forward Price, Germany, €/MWh, Jan-08 to Jan-12]

Figure 113

Expensive gas cannot displace nuclear on an economical basis for baseload electricity.
Source: SG Cross Asset Research, Bloomberg.

In February 2012, GDF SUEZ confirmed almost all its long-term gas contracts have been reviewed to increase market price indexation to above 25%, to lower oil indexed prices and to shorten price review cycles. We therefore estimate that an additional 1.4 bcm of Russian gas has moved from oil-indexation to spot as GDF SUEZ – Gazprom contract was formerly with 15% spot indexation.

In March 2012, ENI and Gazprom reached an agreement on gas supply contracts. ENI / Gazprom revised both prices and flexibility, but didn't disclose details on the agreement reached. We estimated that they agreed a total c.13% discount (taking into account since 2009 renegotiations) and kept a full oil-indexed formula. With this discount, Gazprom gas is now more competitive than Statoil gas, which faces further renegotiations in 2 12e.

– Mostly spot index

In March 2012, E.ON and Statoil agreed a "structural" solution over price of long term gas. This is a long term answer allowing E.ON not to be in loss-making situation of buying at higher prices than reselling at. As none of the two companies gave detail of the term of the deal, if we assume:

- Statoil/E.ON long term contract is for 15 bcm/y.
- 25% has already been spot index after the 2009 crisis.
- The "structural" fix means that it is now 100% spot index.

It means that this deal moved (since January 2012) an additional 11 bcm from oil-indexation to spot indexation.

Thanks to those recent deals, European gas supply could be 55% oil-index and 45% spot in 2012e.

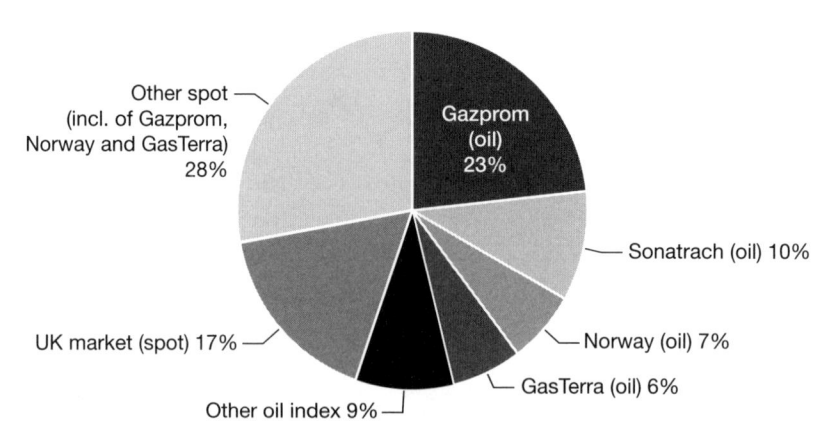

Figure 114

Europe gas supply: 55% oil-linked in 2012e.
Source: SG Cross Asset Research.

26 bcm of contracted gas from GasTerra are also up for renegotiations. Little information is available on those negotiations. But if we assume that those volumes move from 75% oil-indexation to 100% spot, then the split could even be 51% oil-index / 49 % spot.

The closer we go to 50%, the more unstable is the system going to be. So oil-indexation is facing major challenges. The old system were oil-linked long term contracts were signed for both security of demand and security of supply with hub spot trading for additional volumes is facing a step change. Before 2014e, oil-indexation pricing should represent the minority stake in European gas supply. However, long term supply should remain an important tool to fairly share the risk between buyers and producers. In Europe, the rational of oil-indexation has been lost many years ago, so hub pricing makes more sense today.

– *Stake in European utilities*

Gas Natural Fenosa's capital increase was subscribed by Sonatrach in August 2011. Sonatrach acquired 3.85% of Gas Natural Fenosa for €515m in an agreement designed to settle a dispute over the price of long-term gas contracts covering gas supply through the Maghreb-Europe pipeline (9 bcm/y). The dispute was settled in June 2011, when Gas Natural Fenosa agreed to pay $1.9bn for the retroactive revision (1 January 2007 to 31 May 2011) of these contracts.

– Alter contract price formulae after arbitration?

• RWE took Gazprom to arbitration in April 2011 but doesn't expect any resolution until 2013.

• A formal agreement on a gas contract between E.ON and Gazprom was previously reached at the start of 2010, with an estimated 15% spot indexation in the formula. But E.ON would have liked a better deal. With no outcome reached on renewed renegotiations or a definitive fix, E.ON has taken Gazprom to arbitration in 2011. The disconnection between oil and gas prices and the resulting negative gas-oil spread led to considerable margin pressure. The prices of procurement contracts, which are largely oil indexed, are above price levels that can be achieved in E.ON's gas sales business.

• In February 2012, Poland's dominant natural gas company PGNiG filed a suit at the Arbitration Court in Stockholm in an attempt to index more than half of its contracted volumes from Gazprom to spot prices.

The outcome of arbitration is difficult to predict. And during this lengthy process, there may still be concessions on prices to avoid arbitration. But perhaps, arbitrators could decide that long term contracts that used to be oil-index in the 60s, as it was the only price mechanism available then, should now be spot based, as it is the way the majority of gas will be sold in Europe from 2014e. This decision could help the establishment of a single EU gas market and perhaps boost future gas demand...

– Asset swaps?

Joint ventures in power generation could be a fruitful way forward, allowing producers and utilities to share risks and profits and allowing gas to make up the lost baseload electricity caused by the nuclear phase-out.

In June 2012, EDF and Gazprom signed a cooperation agreement on gas power generation in Europe.

– New contracts?

To secure this new gas demand for power generation, producers understand there is a need to agree on new indexes. Selling gas under an electricity index could help utilities to commit capex for building new gas-fired power stations in Europe (that will then operate as tolling plants). But for producers, moving from an oil-index to an electricity-index gas price is a quantum leap even if, at the end of the day, it is still a way to price gas vs. its competing fuel (oil in the 1960s, electricity tomorrow). The ability of Russia to deliver timely physical supply to Europe is not in doubt; its willingness to do so under new indexes (perhaps more volatile) is yet to be ascertained.

– Cancelation of existing contracts?

In April 2012, a French court annulled a gas supply contract from ENI to a French gas-fired production's plant in an attempt to prevent the latter's bankruptcy. This decision followed the implementation of a safeguard procedure which allows French courts to implement measures to improve economic situation of a company. This decision could make buyers' case stronger in arbitration tribunal...

– Long term drawbacks

But moving to a majority gas spot pricing in Europe could further shift the power in the hands of Gazprom at time of high demand if we don't, at the same time, manage to increase domestic production, increased import infrastructure and / or build new storage (to have ability and options to store gas when spot prices are low and withdraw it when spot prices are high).

As we've seen in the UK where Norway has a market power by providing 28% of the local demand, Russia with 24% could also start to use its market power as we've seen recently. In February 2012, several European countries reported reduced flows of Russian gas, as bitter cold caused soaring demand. Gazprom has, we believe, meat its contractual obligations. But with some European contracts having just been renegotiated (higher spot, lower volumes and/or lower flexibility?) this was the perfect time for Gazprom to show to its buyers that it was not willing to provide the expensive swing supplier service at no cost... To guarantee to meet big future swings in European demand, Gazprom is investing in storage, as we will see later. Unless Europe manages to produce enough domestic shale gas it will have to face the dilemma of the power of Gazprom.

Conclusion: After the US Shale Gas Revolution, the 2020 Gas World

CHINA NEEDS TO SECURE EXTRA GAS FROM 2018e...

With consumption growth outpacing that of production, China has become a net gas importer since 2007 and therefore subject to the regional Asian gas price, which is today indexed to oil.

China, which looks set to become a major gas importer has contracted 65 bcm/y from Turkmenistan, 4 bcm/y from Myanmar and up to 46 bcm/y of LNG on a long-term basis (from Australia, Indonesia, Malaysia, Papua New Guinea, Qatar, and from IOCs portfolios (Total, BG, ExxonMobil, Shell)). China is also in negotiation with Gazprom for 68 bcm/y (30 bcm/y via the Altai pipeline into Western China and 38 bcwm/y via the Eastern route), but the price difference between what Gazprom wants (high oil link) and what China is looking for (low oil link) could further delay the signing of a deal.

With domestic unconventional gas and already-contracted LNG and pipe gas, China would not have to rely on Russian gas before 2018e. Thanks to domestic shale gas production, China would only need to sign the 30 bcm/y contract with Gazprom. The other contract (38 bcm/y) would not be needed this side of 2020e. Indeed, judging by the apparent lack of any progress on the gas pricing issue, it seems Beijing is in no particular hurry to sign any Russian contract as the 115 bcm/y already contracted should secure China's gas needs until 2017e.

China's steps to secure its imports reinforce our view that China would first look into any potential US LNG deal before reverting to Gazprom. China can put US LNG and Russian pipe gas into competition. And at the same landed price, we believe it will prefer US LNG... This means that China would first look into any potential deal to acquire US LNG (made from unconventional gas) and then revert to Gazprom. Shale gas could give China pricing power and lead to dual pricing in Asia.

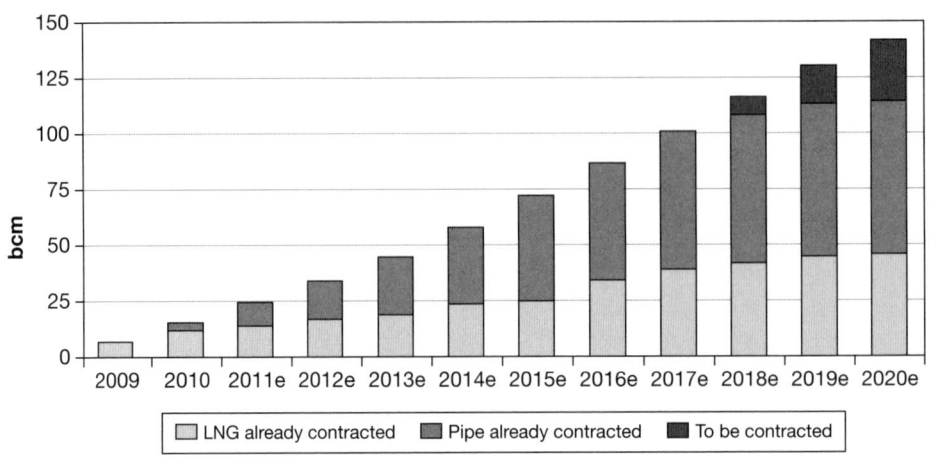

Figure 115

Chinese imports.
Source: SG Cross Asset Research, BP Statistical Review, GIIGNL.

... WHEN US LNG COULD BE AVAILABLE!

Shale gas production has transformed the US gas industry in recent years, boosting production rates and booked reserves of gas. The country has changed from a growing importer into a possible major LNG exporter.

However, opponents of shale gas extraction say fracking chemicals could contaminate water supplies. The major risk for the US not to become a major LNG exporter could materialise if opponents to the fracking technology could prove or convince citizens of the danger of this technology... Only transparency, such as fracfocus.org, can improve acceptability of this technology.

Research projects are targeting new fracking technologies that will use less water. Using less water could: 1/deter opponents' explanations about possible water supplies contamination; and 2/help shale gas production in countries where water is scarce.

The US is facing a choice as actual too low prices do not reflect costs of production. This could lead to scale back in production (less capex, less employment, less taxes)... leading to a higher price at the end of the cycle. With this vicious circle, everybody loses out (producers and consumers). Alternatively, to balance supply and demand, the US could develop its gas demand domestically (gas as a fuel for transportation) but also export LNG. With this virtuous circle, the US will benefit from higher capex (provided by US and non US companies), higher employment and higher taxes for a market price of gas that should always reflect cost of production... US consumers will see a rise in gas prices but this rise should allow price to reflect cost of production. Additionally, by exporting gas to Asia, the US could reduce its trade imbalances.

On top of this, gas flaring is on the rise in the US. The main source of flared gas is coming from shale oil production, where natural gas is produced alongside oil as a by-product. However, the depressed gas prices in North America make it uneconomical to send the gas to market; it's just cheaper to burn it. US gas overabundance would be of much better use supplying foreign markets, instead of wasting a precious resource while harming the environment...

In his State of the Union address, in January 2012, President Obama said "We have a supply of natural gas that can last America nearly 100 years. And my administration will take every possible action to safely develop this energy. Experts believe this will support more than 600,000 jobs by the end of the decade. And I'm requiring all companies that drill for gas on public lands to disclose the chemicals they use. Because America will develop this resource without putting the health and safety of our citizens at risk. The development of natural gas will create jobs and power trucks and factories that are cleaner and cheaper, proving that we don't have to choose between our environment and our economy." This speech reinforces our views that US LNG will materialise as soon as 2016e and that we will see a positive news flow on US and Canada exports (administrative authorisations, FID, new contracts, etc) in the years to come.

North America could export as much as 50 mtpa in 2020e (67.5 bcm is 9% of 2010 North America gas demand). We can split this in 40 mtpa (54 bcm/y) for the US and 10 mtpa (13.5 bcm/y) for Canada.

As 16 mtpa (21.6 bcm/y) have already been contracted from Sabine Pass thanks to Cheniere, this leaves 34 mtpa (45.9 bcm/y) to be contracted. And Asian customers (that have already signed for 7 mtpa (9.4 bcm/y)) would be more than willing to diversify their portfolio away from full oil-indexation by contracting HH gas prices. Asian countries that imported 153 mt (192 bcm) of LNG in 2011, could if they take all incremental LNG and the loss of Indonesian production, receive 264 mt (330 bcm) in 2020e (a 2011-2020e CAGR of 6.2%).

In this case, Asian portfolio would be split: 84% oil-index and 16% HH-index.

As the US will move from a oil importing to a gas exporting country, its focus will shift away from Middle East to China.

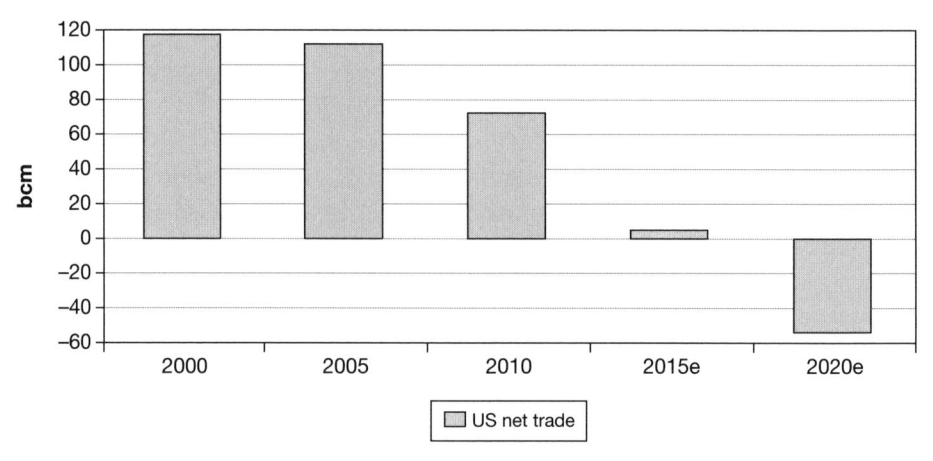

Figure 116

US net gas trade.
Source: BP Statistical Review for historical data.

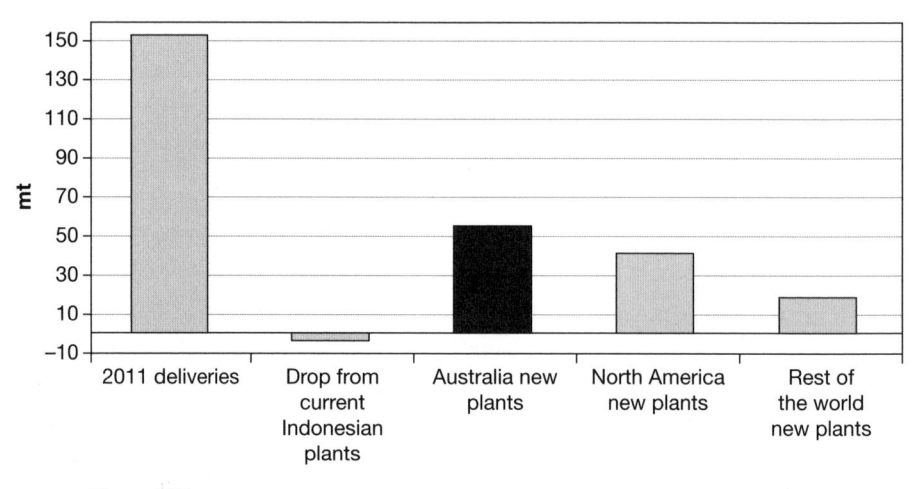

Figure 117

Split of LNG for Asian countries in 2020e vs. 2011.

• Negatively impacting Russia and Australia

Thanks to the shale gas revolution, US has become a low cost energy producer on a global basis. The US could mitigate Russia's power on the international gas scene by delaying its entrance on the Chinese market as China could view US LNG as cheaper and safer than pipe gas from Russia or LNG from the Middle East.

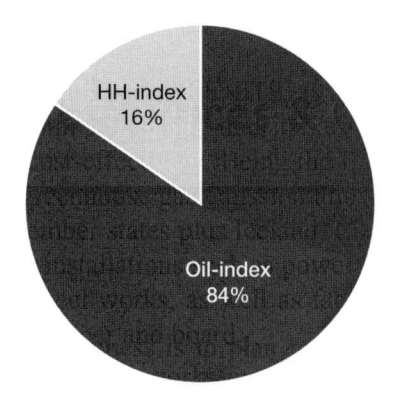

Conclusion: After the US Shale Gas Revolution, the 2020 Gas World

Figure 118

Asian portfolio split in 2020e.

Also, as greenfields projects in Australia are going to be delayed or derailed by the cheaper US LNG exports projects, it is possible that in 2020e, Qatar would still be the number 1 LNG producer, followed by Australia and North America.

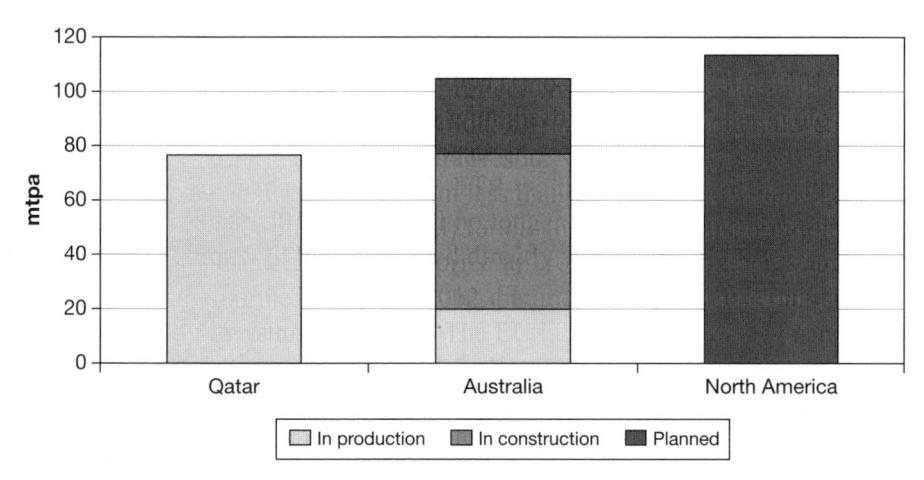

Figure 119

Qatar, Australia or North America: who could be first LNG producer in 2020e?
Source: SG Cross Asset Research.

By directly sourcing US LNG priced under an HH formula, Asian customers (like Gail or Kogas and possibly others) are cutting the "middle man", the LNG aggregator. And if the US becomes a major LNG producer as we believe, then this

Figure 120

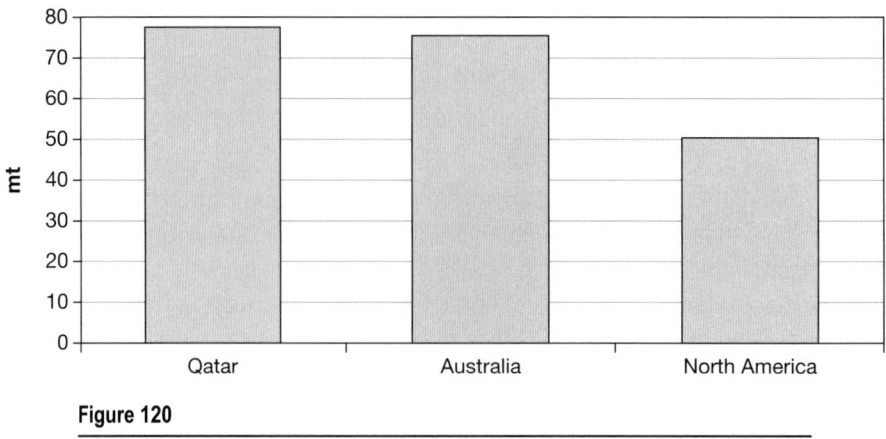

2020e LNG production: 3 first producers.

change in business model could start to reduce oil-indexation in Asia, as we are seeing in Europe…

The US could be the cheapest gas market by the end of the decade; other markets will be linked via the cost of arbitrage (liquefaction, transport and regasification).

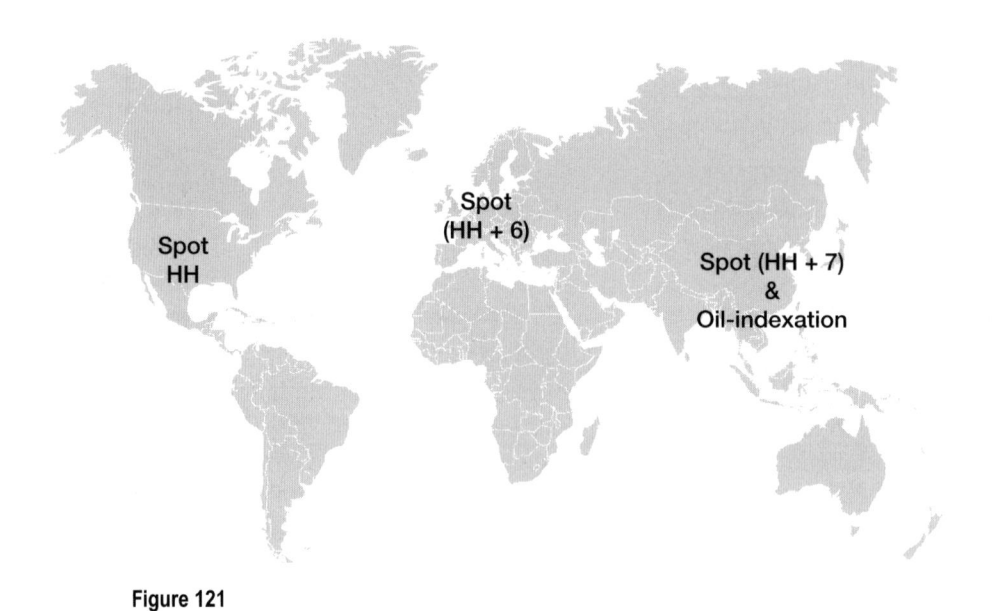

Figure 121

Overview of gas prices in 2020e (with estimated spreads in $/MBtu).

– China to select the winner…

To supply growing markets, the major resource holder, Russia, is now in direct competition with the major gas producer, the US. China has the potential not only to select the winner but also to decide the pricing principle for all Asian buyers in 2020e. As China is a new and growing gas importer and has a lower price tolerance than historical Asian buyers (Japan and South Korea), it is highly possible that, against basic geography, China selects waterborne US LNG vs. close Russian pipe gas, to achieve lower import price.

– … and Singapore to become the Asian LNG hub

Singapore's LNG terminal should start operations in Q2 13e. The initial capacity will be 3.5 mtpa, which will increase to 6 mtpa by the end of 2013e when the third LNG storage tank is brought into service (the terminal can accommodate 7 storage tanks when fully expanded). The main role of the terminal is to supply the Singapore market, but the functionality of the terminal to operate as a trading hub has been built into the design and the commercial arrangements. Singapore will help balance supply and demand in Asia by allowing arbitrages of LNG cargoes. Another hub could later emerge in China (Shangai) but we have discounted this on this side of 2020e.

GAZPROM IN THE DRIVING SEAT IN EUROPE

Europe, so risk adverse that it won't be able to take any decision regarding shale gas production on this side of 2020e, should see its power fading on the energy scene and would rely more on Russia. Gas geopolitics could tighten Russia's stronghold on Europe, on one side, and create a flourishing North America-Asian trade… Russia should in 2020e continue to be the major gas exporter and EU's the major gas importer…

Russia's dominance over European gas is not coming to an end. With advances in drilling techniques, such as horizontal drilling, that have led to new gas discoveries, European gas output could grow but fear of pollution are leading European governments to ban this technology. This fear could lead Europe into a prolonged poor economic growth as it will have to pay higher gas prices than in the US. A back-of-the-envelope calculation shows that the total European gas demand for 2012e (485 bcm) at 460 $/1,000 cm for the oil-linked and 315 $/1,000 cm for the spot gives an average price in Europe of 400 $/1,000 cm, compared with 80 $/1,000 cm for the US. For Europe, the "overprice" in terms of the bill is therefore 325 $/1,000 cm or $155bn for 2012e (0.9% of European GDP)!

Gazprom with 24% market share in Europe will stay the major gas supplier. LNG accounted for 14% of EU consumption in 2011. With regas load factor in Europe at 46%, those facilities could be used more. But with the Fukushima disaster and LNG cargoes being diverted East, Europe LNG imports will decline until 2016e, allowing Gazprom to fill in the gap and increase its market share. Even the prospect of significant North America LNG exports in the second half of the decade shouldn't pose a risk to Gazprom's strategy in Europe as Asia should pre-empt them.

And while it could be in Europe's interest to explore alternative sources for its gas needs, it is uncertain whether Europe as a whole is willing to replace a significant level of imports of Russian gas.

Gazprom is building import infrastructure (Nord Stream and possibly South Stream) and storage in Europe, as it sees this as strategic assets going forward.

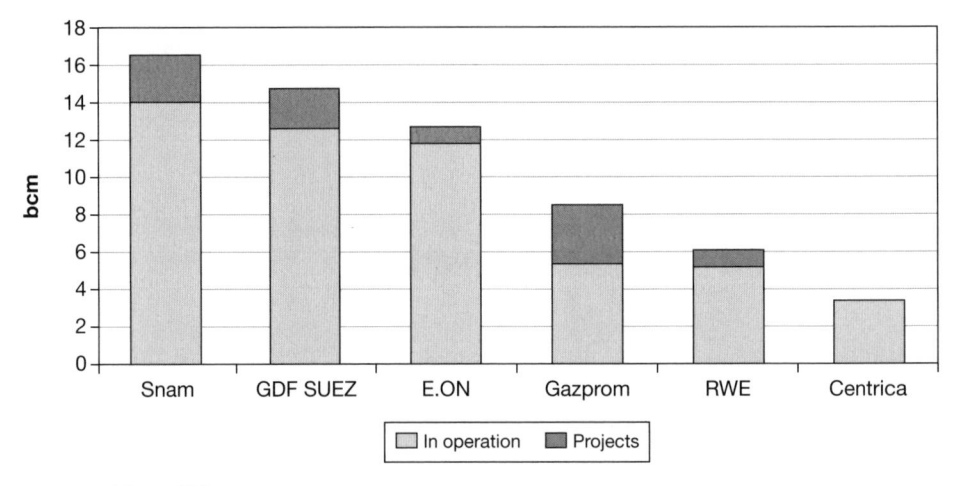

Figure 122

Major storage operators in Europe (working capacity on an equity basis).
Source: SG Cross Asset Research.

Gazprom understands that Europe is going to continue to be its main major market for the foreseeable future, hence the recent contracts renegotiations and price reviews. Gazprom is interested in maximising its gas rent which is both a function of volumes and prices. Thanks to its in-depth knowledge of the European gas market, Gazprom will continue to fine tune its pricing policy to make sure it keeps its market share. Gas will continue to play a significant role in Europe-Russia at least for this decade as Europe's security of supply and Russia's security of demand would continue to be interlinked. It remains to be seen if Europe as a whole will deal with Russia or if it will continue to be on an individual member state level.

Gazprom's old target to diversify away from Europe into the US has failed due to the shale gas revolution. The actual target to rely both on Asia and Europe is challenging and will take time. For this decade, Gazprom will rely exclusively on Europe for its rent maximisation, and thus its biggest source of foreign income. And even if Europe market moves to near full spot indexation, if it is linked to the US via the costs of LNG exports, Europe price could remain 6 $/MBtu (cost of liquefaction, shipping and regasification) above HH. With a minimum price with zero margin at 5.9 $/MBtu as we've discussed earlier for Gazprom gas into Europe in 2020e, this system (as long as Europe doesn't go for unconventional gas) guarantees a market and a margin for Gazprom. The profit will be at least 70% of the HH price (whatever the price is)! The remaining 30% will be taken by the Russian State as export duty (on top of the MET and the 30% taken on the 6 $/MBtu floor). In short, the liquid US market will guarantee minimum profit for Gazprom and revenues for the Russian state!

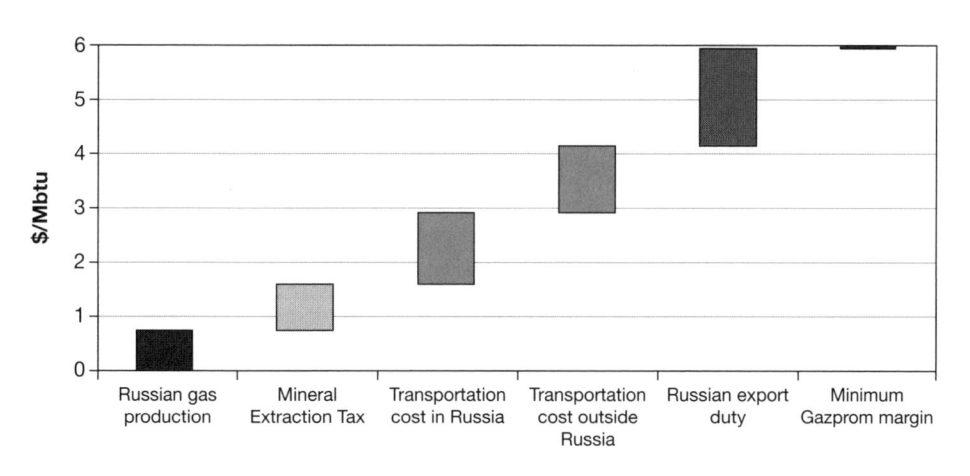

Figure 123

HH price will guarantee Gazprom European rent in 2020e!

THE QUESTION OF ENERGY DEPENDENCY

Until the shale gas revolution net importers were bond to become more and more energy dependant. The shale gas revolution changed this forever dependency paradigm and is offering an alternative… US has chosen to reduce its dependency on foreign (oil &) gas. China will use this new technology to mitigate its growing dependency. Only Western Europe (excluding Poland) has, so far, chosen to avoid this technology and to keep its growing dependency on gas importers. EU gas proven reserves, which have decline 4.4% on a CAGR in 2000-2010, can only grow if Europe decides to go for shale gas. And European gas market will never be fully functioning without enough domestic shale production.

• The Remaining Question of European Oil Dependency

For the first time, in human history, a part of society, which is refusing technology changes, pushes for the full group (Europe) to prefer the status quo! But can this lobby apply its success in refusing technology changes when the question of energy dependency will move from gas to oil?

After the shale gas revolution, we are witnessing a shale oil revolution in the US. Western European countries can oppose shale gas production as they view Russia as a safe gas provider. Hopefully, those lobbyist groups would have more difficulties in opposing shale oil production as the major oil exporter (Saudi Arabia) is neither a democratic regime nor a place where men and women are treated equally. It is going to be much more challenging for those lobby groups to oppose shale oil production to favour non democratic regimes…Saddam's oil, Kaddafi's oil and Ahmadinejad's oil have been viewed, for too long, as needed oil. But now thanks to unconventional oil production we can review our dependency on the Saud family, unless the regime there changes fast for the better…

Units and Conversions

Gas or Natural Gas is mainly methane.

Gas is measured in bcm (billion cubic metres (Gm^3 can also be used)), mcm (million cubic metres), m^3 (cubic metres (cm can also be used)) or cubic feet (cf).

Energy is measured in TWh (Tera Watt hours), GWh (Giga Watt hours), MWh (Mega Watt hours), MBtu (million British thermal units) or th (therms). In Australia, a Petajoules equivalent is used (PJe).

Consumption, flows and demand are in bcm/y (billion cubic meter per year), mcm/d (million cubic metres per day), TWh/y, GWh/d.

Liquefied Natural Gas (LNG) figures are in million tonnes (mt), LNG m^3 (liquid cubic metres) or, transformed into gas, mcm. Capacity and flows of LNG are in mtpa (million tonnes per annum).

Gas is also sometimes expressed in barrels of oil equivalent (boe).

In the US, Henry Hub is the geographic sales point for natural gas on which prices (HH) and futures market contracts are focused. It's in Louisiana.

Prices are in $/MBtu in the US and Asia, in p/th at the National Balancing Point (NBP, UK) and Zeebrugge (Belgium), in $/1,000m^3 for Russian gas sold in continental Europe (linked to oil) and in €/MWh on Continental European hubs.

1 bcm = 1,000 mcm; 1 mcm = 1,000,000 cm

The Russian cm is less than the European cm (the conversion factor is 0.933 to adjust to Russian standard terms and conditions).

1 m^3 LNG (at -162°C) = c. 600 m^3 gas (at 15°C)
1 mt LNG = c. 1.35 bcm gas
1 MBtu = 10 th = 0.293 MWh
1 m^3 gas = c. 11 kWh
1,000 m^3 = 35.31 MBtu
1 bcf/d = c. 10.34 bcm/y
1 bcf = c. 1 TBtu
1 TJ = 278 MWh
Equivalent oil: 1 barrel of oil = 5.8 MBtu; 1Mboe = 5.83 PJe

The full oil parity price for gas is 17 $/MBtu, if oil is at 100 $/b. A generic formula for a long-term Asian LNG supply contract would give 16 $/MBtu when oil is at 100 $/b.

Organizations and Data Providers

BP Statistical Review: the annual review of world energy published by BP. *bp.com/statisticalreview*

CIS: Commonwealth of Independent States, a regional organization whose participating countries are former Soviet Republics, formed during the breakup of the Soviet Union. The Customs Union between Belarus, Kazakhstan, and Russia came into existence in 2010; customs borders between each other have been removed.

EU: European Union consists of 27 Member States. The actual EU-27 Member States are (with year of entry): Austria (1995), Belgium (1952), Bulgaria (2007), Cyprus (2004), Czech Republic (2004), Denmark (1973), Estonia (2004), Finland (1995), France (1952), Germany (1952), Greece (1981), Hungary (2004), Ireland (1973), Italy (1952), Latvia (2004), Lithuania (2004), Luxembourg (1952), Malta (2004), Netherlands (1952), Poland (2004), Portugal (1986), Romania (2007), Slovakia (2004), Slovenia (2004), Spain (1986), Sweden (1995) and United Kingdom (1973). Croatia should be the 28th Member State to join on 1 July 2013. *europa.eu*

Europe: As gas data is limited, Europe in this book refers to European states that are IEA members: Austria, Belgium, Czech Republic, Denmark, Finland, France, Germany, Greece, Hungary, Ireland, Italy, Luxembourg, Netherlands, Norway, Poland, Portugal, Slovakia, Spain, Sweden, Switzerland and United Kingdom. Turkey is excluded as its gas grid is not connected to the rest of Europe.

Eurogas: European organisation composed of 50 members from 27 countries amongst which 33 natural gas companies. Eurogas provides data on the European gas market. *eurogas.org*

FERC: The US Federal Energy Regulatory Commission is an independent agency that regulates the interstate transmission of electricity, natural gas and oil. FERC also reviews proposals to build LNG terminals and interstate natural gas pipelines as well as licensing hydropower projects. *ferc.gov*

GECF: The Gas Exporting Countries Forum was set up with the objective to increase the level of coordination and strengthen the collaboration between the 12 member States that are gas producers (Algeria, Bolivia, Egypt, Equatorial Guinea, Iran, Libya, Nigeria, Oman, Qatar, Russia, Trinidad and Tobago and Venezuela). Kazakhstan, Norway, and Netherlands are Observer Members. *gecf.org*

GIE: Gas Infrastructure Europe is an association representing the sole interest of the infrastructure industry in the natural gas business such as Transmission System Operators, Storage System Operators and LNG Terminal Operators. On top of contributing to a safe and reliable European transmission system, GIE provides on its website detailed maps and data on storage (capacity, projects and daily aggregate inventory level), LNG (capacity and projects) and transmission (capacity). *gie.eu*

GIIGNL: Groupe International des Importateurs de GNL (International Group of LNG importers) is a non-profit organisation that studies and promotes the development of activities related to LNG, in particular purchasing, processing, importing, transporting, handling, regasification and various uses of LNG. GIIGNL has a worldwide focus and its membership is composed of nearly all companies in the world active in the import and regasification of LNG (68 member companies from 21 countries). GIIGNL produces an extensive (and free) annual review of the LNG industry. *giignl.org*

IEA: The International Energy Agency is an autonomous organisation linked with the OECD which works to ensure reliable, affordable and clean energy for its 28 member countries and beyond. Founded in response to the 1973/4 oil crisis, the IEA's initial role was to help countries co-ordinate a collective response to major disruptions in oil supply through the release of emergency oil stocks to the markets. Today, the IEA's main areas of focus are: energy security, economic development, environmental awareness and worldwide engagement. *iea.org*

Kyoto Protocol: An international agreement linked to the United Nations Framework Convention on Climate Change. The major feature of the Kyoto Protocol is that it sets binding targets for 37 industrialized countries and the European community for reducing greenhouse gas (GHG) emissions. These amounts to an average of 5% against 1990 levels over the 2008-2012 period. The Protocol was adopted in Kyoto, Japan, on 11 December 1997 and entered into force on 16 February 2005. Canada notified on 15 December 2011 that it had decided to withdraw from the Kyoto Protocol (to become effective on 15 December 2012).

NPD: Norwegian Petroleum Directorate is a governmental specialist directorate and administrative body that reports to the Norwegian Ministry of Petroleum and Energy. The NPD is responsible for data from the Norwegian continental shelf. *npd.no*

OECD: The Organization for Economic Co-operation and Development is to promote policies that will improve the economic and social well-being of people around the world. The 34 member countries include many of the world's most advanced countries but also emerging countries like Mexico, Chile and Turkey. *oecd.org*

OPEC: Organization of the Petroleum Exporting Countries is an intergovernmental organization of 12 oil-exporting developing nations that coordinates the petroleum policies of its member countries. Algeria, Angola, Ecuador, Iran, Iraq, Kuwait, Libya, Nigeria, Qatar, Saudi Arabia, United Arab Emirates, Venezuela. *opec.org*

SEC: The mission of the US Securities and Exchange Commission is to protect investors, maintain fair, orderly, and efficient markets, and facilitate capital formation. *sec.gov*

UN: The United Nations is an international organization founded in 1945 after the Second World War by 51 countries committed to maintaining international peace and security, developing friendly relations among nations and promoting social progress, better living standards and human rights. Due to its unique international character, and the powers vested in its founding Charter, the Organization can take action on a wide range of issues, and provide a forum for its 193 Member States to express their views, through the General Assembly, the Security Council and other bodies and committees. *un.org*

UEA: The United Arab Emirates is a federation of seven autonomous emirates: Abu Dhabi, Dubai, Ajman, Al Fujayrah, Sharjah, Ras al Khaymah, and Quwayn. Only Abu Dhabi produces LNG.

UK: United Kingdom

UK DECC: UK Department of Energy & Climate Change provides UK energy data. *decc.gov.uk*

US: United States of America

US DoE: US Department of Energy. The US Energy Information Administration is the statistical and analytical agency within the US DoE. EIA collects, analyzes, and disseminates independent and impartial energy information to promote sound policymaking, efficient markets, and public understanding of energy and its interaction with the economy and the environment. By law, EIA data, analyses and forecasts are independent of approval by any other officer or employee of the US Government. To avoid confusion with the IEA, data from the EIA is referred in this book under US DoE. *eia.gov*

Glossary

Associated Natural Gas: Natural gas produced from a well in conjunction with oil or liquids.

Brent: Crude oil produced from the North Sea. Brent is a traded on the ICE. It is a commonly used index in long term gas contracts in Continental Europe.

CAGR: Compound Annual Growth Rate.

Capex: Capital Expenditure. Funds used by a company to acquire or upgrade physical assets fields or equipments.

CBM: Coal Bed Methane.

CNG: Compressed Natural Gas used in vehicles.

CO_2: Carbon dioxide.

Conventional: Hydrocarbon reservoirs which don't require extraordinary means to be produce.

Developmental or exploitation well: Wells drilled within the proved area of an oil or natural gas reservoir known to be productive.

Dry Natural Gas: Natural gas that is produced from a 'gas well' rather than from a well with oil or liquids.

Dutch curse: the economic phenomenon in which the revenues from natural resource exports damage the nation's other productive economic sectors. This curse first became apparent after the discovery of the Groningen field. With the Netherlands' focus primarily on the new gas exports in the 1960s, the Dutch currency grew at a very quick rate which harmed the country's ability to export other products and the Netherlands experienced recession. This process has then been witnessed in multiple resource riche countries.

E&P: Exploration and Production.

FID: Final Investment Decision.

Fracking: See Hydraulic fracturing.

GDP: Gross Domestic Product.

GTL: Gas To Liquids, a process to convert gas into high quality liquid synthetic fuels.

Horizontal drilling: Technique used in certain formations where a well is drilled vertically to a certain depth and then drilled horizontally.

Hydraulic fracturing: Process for extracting gas (and oil) by injecting pressurized water with chemicals to stimulate production wells.

ICE: IntercontinentalExchange, trading place located in London (UK).

IOC: International Oil Company.

JCC: Japanese Crude Cocktail. JCC is the average price of customs-cleared crude oil imports into Japan as reported in customs statistics. It is a commonly used index in long term LNG contracts in Asia.

Liquefaction train: Industrial unit transforming gas into LNG by cooling it down to -162°C.

LNG: Liquefied Natural Gas.

Merchant plant: Buys the input and sells the output in the competitive wholesale market. The investor is taking the commodity risk, with its upside as well as downside regarding revenues and profits.

NGL: Natural Gas Liquids (ethane, butanes and pentanes).

NOC: National Oil Company.

NYMEX: New York Mercantile Exchange, trading place located in New York (US).

Possible Reserves: Defined as "having a chance of being developed under favourable circumstances". Industry refers to this as having a 10% certainty of being produced and known as 3P (Proven plus Probable plus Possible). 3P reserves are 2.3x higher than 1P reserves.

Probable Reserves: Defined as "Reasonably Probable" of being produced using current technology at current prices, with current commercial terms and government consent. Industry refers to this as having a 50% certainty of being produced and known as 2P (Proven plus Probable). This number is used by financial analysts to assess the valuation of a given field or company.

Productive wells: Wells that produce commercial quantities of hydrocarbons.

Proven reserves: Quantities that geological and engineering information indicates with "reasonable certainty" can be recovered in the future from known reservoirs under existing economic and operating conditions. Also known in the industry as 1P or having a 90% certainty of being produced.

Reservoir: A porous and permeable underground formation containing a natural accumulation of producible crude oil and/or natural gas that is confined by impermeable rock or water barriers and is individual and separate from other reservoirs.

Reserves-to-production (R/P) ratio: If the proven reserves remaining at the end of any year are divided by the production in that year, the result is the length of time that those remaining proven reserves would last if production was to continue at that rate. It is just a theoretical number that is used by analysts to rank countries and/or companies reserves.

Resources: Hydrocarbons which may or may not be produced in the future. When the relevant government body gives a production licence which enables the field to be developed, reserves can be formally booked.

Tolling plant: Does not require the owner to purchase the input material or to sell the output product. A fixed charge is set for the running of the plant, allowing the investor to have long-term secured revenues (normally with a low margin).

Spread: Difference between 2 prices. It can be a location spread (prices in 2 areas), timing spread (between 2 different periods) or to calculate profitability (between 2 products).

Unconventional: Upstream activities that require horizontal wells and hydraulic fracturing.

WTI: West Texas Intermediate: Benchmark crude oil price delivered in Cushing, Oklahoma, US.

List of Figures, Maps and Tables

Figure 1	Split of gas proven reserves between major countries at the end of 2010.	5
Figure 2	Split of oil proven reserves between major countries at the end of 2010.	6
Figure 3	2000-2010 CAGR of gas proven reserves in selected countries.	7
Figure 4	2010 gas production split by countries.	8
Figure 5	US has overtaken Russia as the first worldwide gas producer since 2009.	8
Figure 6	Split of US yearly gas production.	9
Figure 7	2010 split of production by companies.	10
Figure 8	2010 Major gas consumers: split by countries.	11
Figure 9	2000-2010 CAGR of gas consumption in selected countries.	11
Figure 10	2010 Net exporters.	12
Figure 11	GECF: 2010 net exports.	13
Figure 12	GECF vs. OPEC: OPEC wins!	13
Figure 13	2010 Net importers.	14
Figure 14	US natural gas production and consumption.	14
Figure 15	UK natural gas production and consumption.	15
Figure 16	China natural gas production and consumption.	16
Figure 17	China net gas imports.	16
Figure 18	Trade grows faster than world consumption.	17
Figure 19	Split between pipe and LNG imports vs. world consumption.	19
Figure 20	From upstream to final customer.	24
Figure 21	Share of 2010 worldwide gas production.	24
Figure 22	Split between LNG and pipe exports on a worldwide basis (2010).	25
Figure 23	2011 World LNG vs. Gazprom production.	25
Figure 24	Top 8 LNG exporters in 2011.	26
Figure 25	Evolution of LNG production of the 2011 top producers.	27

Figure 26	Annual growth in LNG supply.	27
Figure 27	Country LNG load factor in 2011.	28
Figure 28	Historical liquefaction load factor.	29
Figure 29	New liquefaction capacity 2010-2014e.	31
Figure 30	China LNG imports.	32
Figure 31	Worldwide liquefaction capacity and 2011 production vs. regas capacity.	33
Figure 32	Top 8 LNG importers in 2011.	33
Figure 33	European regas capacity by country.	35
Figure 34	Italian split of gas demand after the opening of a new regas terminal.	36
Figure 35	Medgaz should further displace LNG out of Spain.	36
Figure 36	2011 vs. 2010 quarterly variations of Medgaz pipe gas and LNG imports in Spain.	37
Figure 37	UK LNG imports.	38
Figure 38	Gas in primary energy mix for major gas producers.	40
Figure 39	Gas in primary energy mix for major gas consumers.	41
Figure 40	Gas in primary energy mix for major transit states.	41
Figure 41	Max month / Min month in major OECD countries.	42
Figure 42	Max month / Min month in major European countries.	43
Figure 43	UK associated and dry monthly gas production.	44
Figure 44	Split of annual UK gas production.	44
Figure 45	Gazprom daily output.	45
Figure 46	Gazprom monthly exports into Europe.	46
Figure 47	Norway monthly production.	46
Figure 48	Swing factor.	47
Figure 49	Typical storage figures.	49
Figure 50	Split of worldwide storage capacity.	50
Figure 51	Storage capacity for the 3 major consumers in 2010.	50
Figure 52	EU countries with major storage capacities in 2011.	51
Figure 53	How to balance Europe seasonal gas demand in 2012?	52
Figure 54	Seasonal balance of the UK market in 2000.	52
Figure 55	UK seasonal production and demand.	53
Figure 56	Seasonal balance of the UK market in 2010.	53
Figure 57	Gazprom total storage worldwide capacity amounts to 71 bcm.	54
Figure 58	Split of storage capacity between EU, Ukraine and Russia.	55
Figure 59	Demand and storage capacity in EU and Ukraine in 2010.	55

List of Figures, Maps and Tables

Figure 60	Overview of 2012 gas prices.	58
Figure 61	NBP vs. HH.	59
Figure 62	NBP gas, HH gas and Brent oil prices.	61
Figure 63	Brent vs. specific deliveries.	62
Figure 64	NBP vs. specific deliveries.	62
Figure 65	Cost split of Russian gas for Europe.	63
Figure 66	Cost split of Russian gas for domestic market.	63
Figure 67	$18.5bn if Ukraine takes all the rent.	64
Figure 68	Gas trade movements in and out of the UK in 2010.	66
Figure 69	Split of EU 2011 consumption.	67
Figure 70	Europe gas supply: 58% oil-linked in 2011.	68
Figure 71	Major hubs in Europe.	68
Figure 72	How the Interconnector balances UK and Continental markets.	69
Figure 73	Churn ratio on different gas hubs.	71
Figure 74	NBP way ahead of competition…	72
Figure 75	… but TTF is gaining momentum!	72
Figure 76	Forward NBP curve on 1st September of each year between 2003 and 2011.	73
Figure 77	Winter-Summer spread on 1st September of each year between 2003 and 2011.	74
Figure 78	Costs along the LNG chain.	75
Figure 79	East or West? A $3.1m question.	77
Figure 80	Overview of electricity generation from fossil fuels.	79
Figure 81	UK Next Season Clean Spreads.	80
Figure 82	2010 split of CO_2 emissions between OECD and non-OECD	88
Figure 83	2010 top CO_2 polluters.	88
Figure 84	Gas demand for electric generation in Spain.	91
Figure 85	Split of European storage: TPA and non-TPA.	94
Figure 86	US net imports.	98
Figure 87	US yearly gas production.	99
Figure 88	US shale gas changed worldwide ranking.	99
Figure 89	Estimated shale resources in the US.	102
Figure 90	Canada net exports.	102
Figure 91	Major shale gas holders.	103
Figure 92	Shale gas resources vs. proven gas reserves.	104
Figure 93	Shale gas recoverable resources are just estimates.	104

164   *After the US Shale Gas Revolution*

Figure 94	Poland or France, first in Europe?... 107
Figure 95	Incentivised take-or-pay allows the producer to sell extra volume at a discount.. 108
Figure 96	Minimum European price for Gazprom in 2012: 160 $/1,000 cm.. 109
Figure 97	Russian cost of production and Mineral extraction tax for Gazprom's gas... 109
Figure 98	Minimum European price for Gazprom in 2020e: 210 $/1,000 cm. 110
Figure 99	Qatar vs. Australia: LNG capacity... 114
Figure 100	Disclosed capex of LNG projects. 115
Figure 101	CBM in Russia, just profitable for Europe, not for domestic market. 116
Figure 102	2010 split of oil consumption... 126
Figure 103	2010 split of gas consumption.. 126
Figure 104	China consumption... 127
Figure 105	2000-2010 gas demand CAGR in six major EU countries and the EU... 130
Figure 106	Europe gas demand. 130
Figure 107	Split between winter and summer gas consumption in Europe. 131
Figure 108	Europe gas demand assuming a 15 bcm loss from industrials........ 131
Figure 109	German Next Calendar Clean Spreads... 133
Figure 110	ENI Russian Take or Pay obligations. 134
Figure 111	Oil index vs. spot prices... 135
Figure 112	The conundrum: oil-linked prices or renewed dash for gas? 136
Figure 113	Expensive gas cannot displace nuclear on an economical basis for baseload electricity... 137
Figure 114	Europe gas supply: 55% oil-linked in 2012e. 138
Figure 115	Chinese imports.. 142
Figure 116	US net gas trade. 144
Figure 117	Split of LNG for Asian countries in 2020e vs. 2011...................... 144
Figure 118	Asian portfolio split in 2020e... 145
Figure 119	Qatar, Australia or North America: who could be first LNG producer in 2020e?... 145
Figure 120	2020e LNG production: 3 first producers. 146
Figure 121	Overview of gas prices in 2020e (with estimated spreads in $/MBtu).. 146
Figure 122	Major storage operators in Europe (working capacity on an equity basis). 148
Figure 123	HH price will guarantee Gazprom European rent in 2020e!........... 149

Map 1	From East of Caspian to China	23
Map 2	The future supply growth?	118
Table 1	Structure of the Qatari LNG production	26
Table 2	LNG projects up to 2020e	29
Table 3	European hubs competition	70
Table 4	Arbitrage costs	76
Table 5	Comparison between different fuels to generate electricity	83
Table 6	Different TPA regimes in Continental Europe	94

Achevé d'imprimer en janvier 2013 par EMD S.A.S. – 53110 Lassay-les-Châteaux
N° d'impression : 27760 – Dépôt légal : août 2012

Imprimé en France